Live Fire Testing of the F-22

Committee on the Study of Live Fire
Survivability Testing of the F-22 Aircraft
Commission on Engineering and Technical Systems
National Research Council

NATIONAL ACADEMY PRESS
Washington, D.C. 1995

NOTICE: The project that is the subject of this report was approved by the Governing Board of the National Research Council, whose members are drawn from the councils of the National Academy of Sciences, the National Academy of Engineering, and the Institute of Medicine. The members of the committee responsible for the report were chosen for their special competencies and with regard for appropriate balance.

This report has been reviewed by a group other than the authors according to procedures approved by a Report Review Committee consisting of members of the National Academy of Sciences, the National Academy of Engineering, and the Institute of Medicine.

The National Academy of Sciences is a private, nonprofit, self-perpetuating society of distinguished scholars engaged in scientific and engineering research, dedicated to the furtherance of science and technology and to their use for the general welfare. Upon the authority of the charter granted to it by the Congress in 1863, the Academy has a mandate that requires it to advise the federal government on scientific and technical matters. Dr. Bruce M. Alberts is president of the National Academy of Sciences.

The National Academy of Engineering was established in 1964, under the charter of the National Academy of Sciences, as a parallel organization of outstanding engineers. It is autonomous in its administration and in the selection of its members, sharing with the National Academy of Sciences the responsibility for advising the federal government. The National Academy of Engineering also sponsors engineering programs aimed at meeting national needs, encourages education and research, and recognizes the superior achievements of engineers. Dr. Harold Liebowitz is president of the National Academy of Engineering.

The Institute of Medicine was established in 1970 by the National Academy of Sciences to secure the services of eminent members of appropriate professions in the examination of policy matters pertaining to the health of the public. The Institute acts under the responsibility given to the National Academy of Sciences by its congressional charter to be an adviser to the federal government and, upon its own initiative, to identify issues of medical care, research, and education. Dr. Kenneth I. Shine is president of the Institute of Medicine.

The National Research Council was organized by the National Academy of Sciences in 1916 to associate the broad community of science and technology with the Academy's purposes of furthering knowledge and advising the federal government. Functioning in accordance with general policies determined by the Academy, the Council has become the principal operating agency of both the National Academy of Sciences and the National Academy of Engineering in providing services to the government, the public, and the scientific and engineering communities. The council is administered jointly by both Academies and the Institute of Medicine. Dr. Bruce M. Alberts and Dr. Harold Liebowitz are chairman and vice chairman, respectively, of the National Research Council.

This is a report of work supported by Contract MDA972-92-C-0028 between the Advanced Research Projects Agency (ARPA) and the National Academy of Sciences.

Library of Congress Catalog Card Number 95-70169
International Standard Book Number 0-309-05333-1
Additional copies are available for sale from:
National Academy Press
Box 285
2101 Constitution Avenue, N.W.
Washington, D.C. 20055
800-624-6242
202-334-3313 (in the Washington Metropolitan Area)

Copyright 1995 by the National Academy of Sciences. All rights reserved.

Cover photo courtesy of F-22 System Program Office.

COMMITTEE ON THE STUDY OF LIVE FIRE SURVIVABILITY TESTING OF THE F-22 AIRCRAFT

JULIAN DAVIDSON, *Chair*, Booz•Allen & Hamilton, Huntsville, Alabama
DALE B. ATKINSON, Consultant, Springfield, Virginia
JOHN R. BODE, Sandia National Laboratories, Albuquerque, New Mexico
CHARLES C. CRAWFORD, Jr., Georgia Tech Research Institute, Atlanta
ALAN H. EPSTEIN, Massachusetts Institute of Technology, Cambridge
DELORES M. ETTER, University of Colorado, Boulder
DONALD L. GIADROSICH, Consultant, Destin, Florida
ROBERT M. HILLYER, Scientific Applications International Corporation, San Diego, California
ROBERT G. LOEWY, Georgia Institute of Technology, Atlanta
MILTON A. MARGOLIS, Logistics Management Institute, McLean, Virginia
HARRY L. REED, Jr., Consultant, Aberdeen, Maryland
ALTON D. ROMIG, Jr., Sandia National Laboratories, Albuquerque, New Mexico
CHARLES F. TIFFANY, Boeing Military Airplanes (retired), Tucson, Arizona
LAWRENCE G. ULLYATT, Denver Research Institute, Littleton, Colorado
CYNTHIA A. VOLKERT, AT&T Bell Laboratories, Murray Hill, New Jersey

Commission on Engineering and Technical Systems Liaison

ALTON D. SLAY, Slay Enterprises, Inc., Warrenton, Virginia

Staff

BRUCE A. BRAUN, Director, Division of Military Science and Technology
MICHAEL A. CLARKE, Study Director
JOHN A. HUGHES, Project Assistant
NORMAN M. HALLER, Consultant

COMMISSION ON ENGINEERING AND TECHNICAL SYSTEMS

ALBERT R. C. WESTWOOD, *Chair*, Sandia National Laboratories, Albuquerque, New Mexico
H. KENT BOWEN, Harvard University, Boston, Massachusetts
NAOMI F. COLLINS, NAFSA: Association of International Educators, Washington, D.C.
NANCY R. CONNERY, Consultant, Woolwich, Maine
RICHARD A. CONWAY, Union Carbide Corp., South Charleston, West Virginia
SAMUEL C. FLORMAN, Kriesler Borg Florman Construction Co., Scarsdale, New York
TREVOR O. JONES, Libbey-Owens-Ford Co., Cleveland, Ohio
NANCY G. LEVESON, University of Washington, Seattle
ALTON D. SLAY, Slay Enterprises, Inc., Warrenton, Virginia
JAMES J. SOLBERG, Purdue University, West Lafayette, Indiana
BARRY M. TROST, Stanford University, Stanford, California
GEORGE L. TURIN, Teknekron Corp., Menlo Park, California
WILLIAM C. WEBSTER, University of California, Berkeley
DEBORAH A. WHITEHURST, Arizona Community Foundation, Phoenix, Arizona
ROBERT V. WHITMAN, Massachusetts Institute of Technology, Cambridge
CHARLES E. WILLIAMS, Toll Road Investors Partnership II, Sterling, Virginia

Staff

ARCHIE L. WOOD, Executive Director
DENNIS I. CHAMOT, Associate Executive Director
ROBERT J. KATT, Associate Executive Director

Preface

The Live Fire Test Law (10 U.S.C. 2366) mandates realistic survivability and lethality testing of certain systems or programs. The law defines realistic survivability testing as "testing for vulnerability of the system in combat by firing munitions likely to be encountered in combat . . . at the system configured for combat, with the primary emphasis on testing vulnerability with respect to potential user casualties and taking into equal consideration the susceptibility to attack and combat performance of the system." A provision of the law permits the Secretary of Defense to waive these tests if the Secretary certifies to Congress, before a system enters engineering and manufacturing development, that live fire testing "would be unreasonably expensive and impractical."

The Air Force did not request a waiver and the Secretary of Defense did not waive live fire tests for the F-22 combat aircraft before the program entered engineering and manufacturing development. Instead, the Department of Defense later requested that Congress enact new legislation to permit the Secretary of Defense to grant a retroactive waiver. Proposed legislation to this effect was submitted to Congress in October 1993. Rather than enacting such legislation, Congress requested this study.

Specifically, language contained in the National Defense Authorization Act for fiscal year 1995 charged the Secretary of Defense to ask the National Research Council of the National Academy of Sciences "to conduct a study regarding the desirability of exercising the authority . . . to waive for the F-22 aircraft program the survivability tests required pursuant to [the law] . . . "[1,2] The study's Statement of Task, below, was drawn verbatim from the legislation:

---

[1] National Defense Authorization Act for fiscal year 1995, Conference Report to Accompany S.2182 (p. 43), August 12, 1994. The language requiring this study was enacted into law on October 5, 1994 (Public Law 103-337).

[2] The committee notes here that the 1993 National Research Council report *Vulnerability Assessment of Aircraft: A Review of the Department of Defense Live Fire Test and Evaluation Program*, prepared by the Air Force Studies Board Committee on Weapons Effects on Airborne Systems, was cited by Senator Roth in support of the legislation requiring this study. That report, which discussed the general subject of aircraft vulnerability assessment and the effect of the Live Fire Test Law, provided valuable background that allowed the current committee to focus on the

The report shall contain the following matters:

(1) Conclusions regarding the practicality of full-scale, full-up testing for the F-22 aircraft program.[3]
(2) A discussion of the implications regarding the affordability of the F-22 aircraft program of conducting and of not conducting the survivability tests, including an assessment of the potential life cycle benefits that could be derived from full-scale, full-up live fire testing in comparison to the costs of such testing.
(3) A discussion of what, if any, changes of circumstances affecting the F-22 aircraft program have occurred since completion of the milestone II program review to cause the program manager to request a waiver of the survivability tests for the F-22 aircraft program that was not requested at that time.
(4) The sufficiency of the F-22 aircraft program testing plans to fulfill the same requirements and purposes as are provided in subsection (e)(3) of section 2366 of title 10, United States Code, for realistic survivability testing for purposes of subsection (a)(1)(A) of such section.
(5) Any recommendations regarding survivability testing of the F-22 aircraft program that the Council considers appropriate on the basis of the study.

In response to the legislation and a request from the Department of Defense, the Committee on the Study of Live Fire Survivability Testing of the F-22 Aircraft was formed under the auspices of the National Research Council's Commission on Engineering and Technical Systems and its Division of Military Science and Technology to carry out the study. The committee began to function in December 1994.

The full committee met five times over the course of the study. At the early meetings and on other occasions, the members were briefed by representatives of the Air Force, Navy, Office of the Secretary of Defense, and other government and industry officials on matters relating to the F-22 program and live fire testing. (See Appendix A for a detailed listing of meetings and the persons and organizations who addressed the full committee and its members.) Many relevant documents from various agencies were also received.

---

specific case of the F-22. Excerpts from and references to the previous report appear extensively in this report.

[3] Full-scale, full-up testing would subject an F-22 to live fire in its combat configuration, including on-board ordnance and fuel.

During the study, discussions among many experts, both on the committee and from other organizations, allowed full airing of the issues concerning live fire testing. As a result, the committee believes that its study provided ample opportunity for consideration of all sides of the case involving full-up, full-scale testing of the F-22.

The committee began by reviewing the threat, mission, and operational requirements for the F-22 to understand the kinds of hostile environments that the aircraft might encounter in future conflicts. The requirements for live fire testing and the testing plans and assumptions germane to the proposed waiver were then reviewed.

In making the judgments documented in this report, the committee depended to a large extent on information provided by the Department of Defense and on the previous National Research Council report *Vulnerability Assessment of Aircraft*. There was neither the time nor the resources to develop substantial amounts of new information. The committee scrutinized the information it was given, bringing to bear the considerable experience, knowledge, and expertise of its members. The committee then made its assessments and formulated its conclusions and recommendations.

The committee expresses its sincere appreciation to the many individuals and groups who provided invaluable information and support during this study.

JULIAN DAVIDSON
CHAIRMAN

PREFACE

Contents

	Executive Summary	1
	Principal Findings	2
	Practicality and Cost-Benefit	2
	Sufficiency	3
	Vulnerability Assessment Tools	4
	Conclusions and Recommendations	4
	Desirability of Waiver for the F-22 Tests	5
	Changed Circumstances Since Milestone II	5
	Affordability and Cost-Benefit	5
	Sufficiency of Tests Planned for the F-22	6
	Other Recommendations	9
1	Introduction	11
	Vulnerability in the Context of Overall Survivability	11
	Vulnerability Testing of Aircraft Versus Ground Vehicles	13
	Report Organization	14
	References	15
2	Origin of Testing Requirements	16
	Framework	16
	F-22 Live Fire Testing Requirements	17
	The Live Fire Test Law—Requirements and Historical Interpretations	17
	Recent Live Fire Test Guidelines and Interpretations	19
	Recent Amendment to Waiver Provision of the Live Fire Test Law	21
	Request for a Retroactive F-22 Test Waiver	21
	Position on Vulnerability	23
	Position on Full-Up, Full-Scale Testing	23
	Confusion Over Interpretation of the Law	23
	Help From the Previous Committee	24
	A Lingering Question	25
	F-22 Design Requirements for Vulnerability	25

	Disposition of Prior Committee's Recommendations	27
	Summary	27
	Requirements Background	27
	What Changed to Cause Request for Waiver	28
	F-22 Design Requirements for Vulnerability	28
	Disposition of Previous Recommendations	28
	References	29
3	Practicality, Affordability, and Cost-Benefit	31
	Practicality	31
	Relative Importance of Vulnerability Reduction to F-22 Survivability	31
	Realism in Aircraft Testing	33
	Destructive Versus Nondestructive Testing	36
	Expert Opinion	36
	Affordability	38
	Affordability of Full-Up, Full-Scale Testing	39
	Investment Methodology for F-22 Vulnerability Tests	40
	Cost-Benefit Methodology	40
	Conclusions	41
	References	43
4	Sufficiency of F-22 Testing Plans	44
	F-22 Threat Environment and Its Replication	44
	Overview of the Air Force Vulnerability Assessment Program	47
	Evaluation of the Vulnerability Assessment Program	50
	Structure and Integral Fuel Tanks	51
	Fuel System and Associated Dry Bays	59
	Flight Control and Auxiliary Systems	63
	Weapons Bay and Ordnance	65
	Engines	67
	Flight Crew	69
	Fire Protection Systems	71
	Additional Observations	76
	Conclusions	78
	Adequacy of F-22 Threat Definition and Replication	78
	Overall Sufficiency	78
	Specific Actions	79
	Additional Action	80
	References	81

5	Vulnerability Assessment Tools	83
	Role of Testing, Modeling and Data Bases in Vulnerability Assessment	83
	Documentation	85
	Data Bases	86
	Models	86
	Phenomenological Models	87
	Encounter Models	89
	Models Used by the F-22 System Program Office	89
	Large-Scale Effects	90
	Conclusions	91
	References	92
6	Recommendations	94
	Desirability of Waiver for the F-22 Tests	94
	Cost-Benefit Methodology	94
	Sufficiency of Tests Planned for the F-22	95
	Other Recommendations	97
	Vulnerability Requirements	97
	Vulnerability Assessment Tools	97
	Appendix A: Meetings, Ste Visits, and Discussions	99
	Appendix B: Live Fire Test Law	105
	Appendix C: Department of Defense F-22 Waiver Request	108
	Appendix D: Vulnerability Assessment Process	117

List Of Exhibits

Table 4-1	F-22 Threat Environment	45
Figure 4-1	Locations of the test areas	49
Figure 4-2	Structural configuration	52
Figure 4-3	Materials applications	53
Figure 4-4	Aft boom	54
Figure 4-5	Current wing configuration	56
Figure 4-6	Forward boom A1 fuel tanks	57
Figure 4-7	Fuel system vulnerability testing	60
Figure 4-8	Vulnerability reduction features of the F-22	62
Figure 4-9	Areas of interest for Test 6	74

List of Abbreviations

ACC	Air Combat Command
AIM	Air Intercept Missile
AMAD	Airframe Mounted Auxiliary Drive
API	Armor Piercing Incendiary
APU	Auxiliary Power Unit
COVART	Computation of Vulnerable Areas and Repair Times
DoD	Department of Defense
EMD	Engineering and Manufacturing Development
ESAMS	Enhanced Surface-to-Air Missile Simulation
FASTC	Foreign Aerospace Science and Technology Center
FMECA	Failure Modes Effects and Criticality Analysis
HEI	High-Explosive Incendiary
JDAM	Joint Direct Attack Munition
JTCG/AS	Joint Technical Coordinating Group on Aircraft Survivability
JTCG/ME	Joint Technical Coordinating Group on Munitions Effectiveness
LFT	Live Fire Test
LFT&E	Live Fire Test and Evaluation
NRC	National Research Council
OBIGGS	On-Board Inert Gas Generating System
ORD	Operational Requirements Document
OSD	Office of the Secretary of Defense
PAO	Polyalphaolefin (a cooling fluid)
SPO	System Program Office
STAR	System Threat Assessment Report
SURVIAC	Survivability and Vulnerability Information Analysis Center
TAF	Tactical Air Force
TEMP	Test and Evaluation Master Plan

LIST OF ABBREVIATIONS

Live Fire Testing of the F-22

Executive Summary

The Air Force's newest combat aircraft, the F-22, is designed principally for offensive counter-air missions.[1] Advanced technologies, including stealth, are intended to enable it to penetrate deeply into hostile airspace and shoot down threatening aircraft before they can detect the F-22. If detected, the F-22 has the countermeasures, speed, and maneuverability that should minimize the likelihood of its being hit.

The National Defense Authorization Act for fiscal year 1995 directed the Secretary of Defense to request the National Research Council to study the desirability of waiving the live fire tests that are required by law for the F-22. The Committee on the Study of Live Fire Survivability Testing of the F-22 Aircraft was formed by the National Research Council to conduct the study.

The committee began its work in December 1994. Several data gathering meetings were held in Washington, D.C.; one was at Wright-Patterson Air Force Base, Dayton, Ohio, which is the location of the F-22 System Program Office and the Wright Laboratory, where many Air Force live fire tests are conducted; and one was at the Naval Air Warfare Center located at China Lake, California, where the Navy conducts similar tests. These meetings exposed the committee to the full spectrum of views involving live fire testing of fighter aircraft.

The work of a previous National Research Council committee, reported in the 1993 publication *Vulnerability Assessment of Aircraft: A Review of the Department of Defense Live Fire Test and Evaluation Program* , was invaluable to the present committee. The previous committee confirmed that, unless waived by the Secretary of Defense because of unreasonable expense and impracticality, live

---

[1] According to the Air Combat Command, which is the operational user of the F-22, there are two kinds of counter-air missions. The first, known as offensive counter air, is to "penetrate deep into heavily defended hostile airspace and destroy threat capability." The second, defensive counter air, is to "detect, identify, intercept, and destroy threat aircraft penetrating friendly airspace." (W.S. Hinton, "Threat, Mission, and Operational Requirements for the F-22," presentation to the Committee on the Study of Live Fire Survivability Testing of the F-22 Aircraft, Washington, D.C., December 21, 1994.) For the purpose of this report the committee uses the definition of offensive counter air that involves only air-to-air engagements.

fire tests known as full-up, full-scale tests are required for major weapon systems like the F-22.

Full-up, full-scale tests would subject an F-22 to live fire in its combat configuration, including on-board ordnance and fuel. These tests would use munitions likely to be encountered by the F-22 in combat. There is currently no doubt within the Department of Defense about this requirement or the need for a waiver if such tests are not to be performed.

If the tests are waived, the law permits the Secretary of Defense to allow instead live fire tests against components, subsystems, and subassemblies, together with design analyses, modeling and simulation, and analysis of combat data. Along with any waiver, the Secretary must report how system survivability will be evaluated and assess possible alternatives.

Circumstances associated with a waiver for the F-22 are unique in one respect. Normally, if a waiver is to be granted, the Secretary of Defense must act before a weapon system enters the phase of acquisition known as engineering and manufacturing development. The decision point for this phase is called Milestone II. Milestone II for the F-22 occurred in 1991 without a waiver being requested. In 1993, the Department of Defense, along with developing an alternative test plan, requested a change to the law that would permit the F-22 to be granted a retroactive waiver. That request prompted members of Congress to call for this study.

The remainder of this summary contains (a) an overview of the principal findings that resulted from this study, followed by (b) a more detailed presentation of the committee's major conclusions and recommendations.

PRINCIPAL FINDINGS

The committee's principal findings concern three topics: (1) practicality and cost-benefit issues related to live fire testing of the F-22, (2) sufficiency of the current vulnerability assessment program for the F-22, and (3) vulnerability assessment tools.

Practicality and Cost-Benefit

The committee believes that live fire testing can be conducted at four levels of realism for the intended mission.[2] From lowest to highest, these four levels are (1) hardware simulations or mock-ups of production systems; (2) full-scale

---

[2] In the case of the F-22, the primary mission is offensive counter air, in which the Air Force expects very few encounters to result in enemy missiles or guns being fired at the aircraft.

components, subsystems, or major subassemblies representative of production items; (3) a complete production aircraft not loaded with live ordnance or fuel; and (4) a full-up, full-scale ground test.

Even the highest level does not adequately simulate actual flight conditions, and testing the virtually infinite number of threat-target interactions is out of the question. Destructive testing (as opposed to nondestructive testing) of a complete aircraft at the full-scale level yields extremely low confidence factors due to the small number of trials possible. The opinion of most experts with whom the committee met, and the committee's opinion, is that full-scale testing is much less likely to provide useful information than are component, subsystem, and subassembly tests.

The committee carefully weighed arguments on both sides of the issue involving tests of a complete, full-scale aircraft (e.g., the chances that such full-scale tests will reveal "unknown unknowns"). The committee was persuaded that, for a system like the F-22, a well conceived, incremental build-up of tests that proceed from the component level to the subassembly or large assembly levels made the most sense. The combination of lack of realism in test conditions, the difficulty of obtaining a sufficient number of trials, and expert opinion all support the conclusion that completely realistic, destructive, full-up, full-scale testing of the F-22 is not practical and offers low benefits for the costs.

These views formed the basis of the committee's position regarding the waiver requested for the F-22. The committee's recommendation on the waiver appears below.

Sufficiency

The committee evaluated the current vulnerability assessment program for the F-22. The Air Force and its contractors have incorporated many features in the F-22 design that will reduce the aircraft's vulnerability (e.g., a largely multiple load path structural design, fuel tank inerting, and much subsystem redundancy). The design is complemented by a strong vulnerability analysis and live fire test program.

However, the committee has some concerns about the program. For example, no live fire tests are currently planned to assess damage done by direct hits on important aft structural members; the current vulnerability specifications do not include the effects of on-board ordnance; and the analysis does not properly account for the flammable properties of the hydraulic and cooling fluids. The committee believes that several additional tests and analyses are needed to strengthen the program. Specific recommended actions are discussed later.

Vulnerability assessment is a complicated matter. As part of its evaluation, the committee carefully considered the need for continued vulnerability testing of the F-22 beyond the currently planned tests that support engineering and manufacturing development and initial production. The committee held detailed discussions with the vulnerability and lethality assessment communities. In particular, the committee considered the Navy's assessment methodology for fighters.

Large assembly testing of variants of the Navy's F-18 has been conducted and is planned, even though the Navy vulnerability assessment team does not expect to discover unanticipated outcomes. It is possible that additional data could be obtained from similar live fire testing of the major parts of an F-22 and result in revisions to the aircraft design to reduce further its vulnerability. The committee recognizes and accepts that test assets may not become available until after production begins and are not likely to influence the production configuration of the counter-air version of the F-22.

The Navy's rationale for larger scale testing was persuasive: something is always learned, vulnerability assessment tools are evaluated and improved, and the test base cannot be allowed to wither. After much deliberation, the committee agreed that the expeditious conduct of tests against large assemblies made sense for the F-22. This fighter will be in the inventory for decades and can be expected to evolve, to include other missions (e.g., air-to-surface) and new configurations. Testing of its large assemblies could also verify techniques for repairing battle damage to the F-22's new composite materials and systems.

Vulnerability Assessment Tools

To be successful, vulnerability assessment requires much mutual support between documentation, data bases, models, and testing. The committee's review of the F-22 vulnerability assessment program indicated that significant improvements are needed in several of the tools (i.e., documentation, data bases, and models) that complement live fire testing. Specific conclusions and recommendations relating to these tools are provided below.

CONCLUSIONS AND RECOMMENDATIONS

The committee's major conclusions and recommendations follow. They are directed at the specific matters in the legislation that requested this study.

The committee's principal recommendation, which appears immediately below, requires action by Congress. The numbered recommendations that follow require action by the Department of Defense. Specific authorization and appropriation by Congress may be necessary to implement some of the numbered recommendations.

Desirability of Waiver for the F-22 Tests

Principal Recommendation. Permit a waiver of the full-up, full-scale, live fire tests required by law for the F-22. The committee believes that such tests are impractical and offer low benefits for the costs.

Recommendation 1. Interpret a waiver as reinforcing the need to conduct robust live fire tests of the F-22 that build incrementally from the component level to the subassembly or large assembly levels. (Recommendations to strengthen the current test program appear below.)

Changed Circumstances Since Milestone II

Nothing specific to the F-22 program appears to have changed since Milestone II in 1991 that would have prompted the Air Force to request a waiver. However, there appears to be a changed view of what is required to fulfill the intent of the live fire test law as it was interpreted within the Department of Defense. Although there was some consideration of a request for a waiver in 1991, it was not until after the 1993 report by the National Research Council that the Air Force appeared to accept fully the fact that a waiver was necessary. The Air Force then instituted the current live fire test program, which does not include a full-up, full-scale test, and the waiver request followed.

Affordability and Cost-Benefit

The committee's conclusions relative to the impracticality and low benefits for the costs of full-up, full-scale, live fire testing were provided above. The committee was also asked to address affordability.

The committee defines an "affordable" activity as one within budgetary constraints or attainable budgets. A full-up, full-scale test of the F-22 would cost approximately $250 million (then-year dollars) above the currently planned program. Thus, to perform the test, either more funding would need to be authorized, or the test would have to displace $250 million of currently funded activities in the program.

There is an argument that even $250 million is only a small percentage (less than 0.5 percent) of total F-22 program costs, and therefore cost should not be a determining factor. The committee was not able to consider fully or challenge the prioritization of items within the current F-22 program budget. However, the committee's judgment (discussed earlier) is that the benefits of full-up, full-scale testing of the F-22 are low relative to the costs. Even if $250 million were provided for additional vulnerability assessment of the F-22, the committee would not support using the funds for full-up, full-scale testing.

The committee concludes that the affordability of live fire tests is not the matter of foremost relevance; cost-benefit is most relevant. Affordability only becomes relevant if the benefits relative to the costs of whatever tests are being considered are commensurate with the benefits relative to the costs of other alternatives.

The committee notes that its judgments regarding the costs and benefits of full-up, full-scale testing were reached in the absence of a mature methodology for assessing benefits relative to costs. The committee reviewed cost-benefit methodologies related to incremental live fire testing; conclusions and recommendations concerning these methodologies are covered next.

Despite recommendations made by the previous National Research Council committee, this committee regrets the lack of more progress in developing a cost-benefit methodology for determining the return on investment of successive levels of live fire tests. The methodologies briefed to the committee during the study are immature and appear to address suboptimal measures of benefit. The committee is leery of reliance on methodologies that use an overly simple construct of the F-22's future to make judgments about how far to go with live fire testing. A broader analytical framework could elevate the importance of reduced F-22 vulnerability over the long term and might enhance the benefits relative to the costs of given levels of testing.

Recommendation 2. Continue Department of Defense efforts to develop viable cost-benefit methodologies for planning the extent of live fire testing. Pursue methodologies to examine cost-benefit issues in the light of frameworks that take a broad view of how the future may develop for weapon systems like the F-22.

Sufficiency of Tests Planned for the F-22

The committee's evaluation of the current vulnerability assessment program considered threat-related assumptions as well as the attendant vulnerabilities and analyses and tests associated with the F-22's major subsystems (i.e., structure and integral fuel tanks, fuel system and dry bays, flight control and auxiliary systems, weapons bay and on-board ordnance, engines, flight crew, and fire protection systems). The major results of the committee's evaluation are summarized below.

The committee accepts the threat environment defined for the current mission of the F-22. Threat replication in the test program is reasonable. However, for some classes of anti-air missile warheads (e.g., annular or focused blast fragmentation), the kill mechanism that involves dense multiple fragment impacts may be important for the F-22.

Recommendation 3. Consider, in future analyses and tests, the kill mechanism that involves dense multiple fragment impacts.

Overall, the committee believes that the vulnerability assessment program for the F-22, given its current counter-air mission, is sufficiently realistic to support the requested waiver. The following specific actions are recommended by the committee to strengthen the program as the F-22 proceeds with engineering and manufacturing development and initial production.

Recommendation 4a. Conduct additional live fire testing to determine the damage that can be expected from a hit in the Frame 6 aft boom attachment area. Determine the most critical shot lines for this testing.

Recommendation 4b. Expand analyses to predict damage sizes and residual strengths of the aft boom, Frame 6, and horizontal tail pivot shafts after being hit by 30mm high-explosive incendiary rounds. Also, determine the risk of aircraft loss should it be found that loss of a horizontal tail is possible.

Recommendation 4c. Conduct further analysis of the aft fuel tank (A-1) prior to the conduct of Test 4D.[3] Focus this analysis on determining the adequacy of the test specimen, with particular emphasis on its ability to simulate accurately the reaction of the entire tank.

Recommendation 4d. Make the operational community fully aware that a fuel ingestion risk to the aircraft exists at a fuel state higher than 60 percent. (This risk arises because the fuel tanks next to the engine inlets are not empty at fuel states above 60 percent; thus, a puncture could lead to fuel ingestion by an engine and potential engine failure.)

Recommendation 4e. Conduct the tests and analyses, proposed by the F-22 System Program Office, on the flammability of coolant and other fluids and the attendant vulnerability of the aircraft.

---

[3] Test 4D was planned for August 1995, when a section of the fuel tank between Frames 5 and 6 was to be filled with water, externally loaded, and hit with a 30mm high-explosive incendiary round.

Recommendation 4f. Undertake the analysis and test, proposed by the System Program Office, of ablative materials in the weapons bay. Also, conduct further analysis of the tradeoffs associated with additional ordnance protection or defensive measures.

Recommendation 4g. Fund the Joint Technical Coordinating Group on Aircraft Survivability and the Joint Live Fire Test Program to assure the completeness of data on the vulnerabilities of on-board ordnance.

Recommendation 4h. Fund the proposed Joint Live Fire testing of F119 engine components to alleviate the paucity of testing against those components.

Recommendation 4i. Emphasize continuing efforts by the F-22 System Program Office and the Joint Technical Coordinating Group on Aircraft Survivability to develop improved methodologies for reducing flight crew vulnerability.

Recommendation 4j. Use the prototype air vehicle fuselage in Test 6A in lieu of a mock-up. (The Air Force has considered using this fuselage for Test 6A, which will examine the synergistic effects of pressurized coolant lines and cooled avionics modules and the adjacent powered electrical wiring in the F-22 forward fuselage lower avionics bays.)

In addition, the committee recommends that the Air Force begin planning for expeditious vulnerability assessment testing of the F-22 similar to that being conducted or planned for variants of the Navy's F-18.

Recommendation 5a. Use large subassemblies from production-representative hardware (e.g., a damaged aircraft or other source) in these tests.

Recommendation 5b. Provide these assets, as soon as they become available, to the vulnerability assessment community for the conduct of live fire tests.

Recommendation 5c. Direct the tests at (a) verifying predictions from the current F-22 live fire test program and the models used, and (b) testing the effects on overall F-22 vulnerability assessment brought about by configuration and mission changes. Also, use the tests to verify techniques for repairing battle damage to the F-22's new composite materials and systems.

Other Recommendations

Vulnerability Requirements

Given the F-22's counter-air mission, the Air Force placed a premium on achieving high survivability through reduced aircraft susceptibility. Reduction of vulnerability in the design, although important, was deemed less significant. The user (Air Combat Command) did not specify quantitative vulnerability requirements for the F-22. Instead, contract specifications for vulnerability were developed by the System Program Office in coordination with the user to establish a basis for design optimization and assess contractor performance.

The existing vulnerability specifications and the live fire test program do not address future missions that the F-22 might be required to perform. The System Program Office (the developer) is currently planning for an air-to-surface mission using ordnance delivered from relatively high altitude. At some point in the future, lower altitude missions could be required.

Recommendation 6. Reexamine expeditiously, for future F-22 missions (e.g., air-to-surface), the balance of requirements among susceptibility, vulnerability, and related performance parameters.

Recommendation 7. Include in operational requirements for any new missions user validation of quantitative vulnerability requirements, and plan new live fire tests as necessary in response to those requirements.

Vulnerability Assessment Tools

The committee reviewed the documentation, data bases, and models that complement live fire testing for the F-22 and other aircraft programs.

Most of the documents (e.g., standards for vulnerability design) were found to be 10 or more years old and in need of updating and improvement.

Recommendation 8. Update and improve expeditiously the various standards, handbooks, and design guides that are important to the aircraft vulnerability community.

Data bases play a distinct role in vulnerability assessment in that they form the institutional memory that bridges specific systems, prevents repetitious testing, and avoids the mistakes of the past. Insufficient component tests have been accomplished to produce confidence in the data bases for existing components and materials; and new composite materials, engines, stealth techniques, and other advances will require additional testing. The committee believes there is considerable need for expanded efforts to improve the data bases.

Recommendation 9. Direct the Joint Technical Coordinating Group on Aircraft Survivability to define and plan a Joint Live Fire Test Program that will, over the next several years, produce sound vulnerability data bases; apply aggressive funding to implement this program.

Modeling is essential because the virtually infinite possibilities for threat-target interactions demand ways to extend the limited number of tests that can be conducted. Experimentation has been the main approach for obtaining data to empirically fit the models. The advanced modeling techniques used by the designers of nuclear weapons and the aerospace and automotive industries have not been exploited by the vulnerability community. For the F-22, the committee judges that a relatively large uncertainty is the ability to model the response of its composite materials. Also, valid large-scale models could provide an efficient means of making sound judgments without the need for expensive and repetitive live fire tests on large subassemblies.

Recommendation 10a. Validate and accredit formally, by the Joint Technical Coordinating Group on Aircraft Survivability and the Joint Technical Coordinating Group on Munitions Effectiveness, the vulnerability assessment models used by the Air Force and other services.

Recommendation 10b. Improve the vulnerability models of the vulnerability community, and adopt these improvements for the F-22.

Recommendation 10c. Explore the application of advanced methodologies currently being used by nuclear weapons designers and other industries.

Recommendation 10d. Focus on ways to understand fully the response of F-22 composite materials to ballistic damage, and develop and exercise analysis tools that can handle large-scale damage effects.

1

Introduction

This chapter sets forth the overall framework within which the committee performed its task. It also furnishes a road map for the rest of the report.

VULNERABILITY IN THE CONTEXT OF OVERALL SURVIVABILITY

The purpose of the Air Force's newest combat aircraft, the F-22, is to conduct counter-air missions. According to Air Combat Command, the operational user of the F-22, there are two kinds of counter-air missions. The first, offensive counter air, is to "penetrate deep into heavily defended hostile airspace and destroy threat capability"; the second, defensive counter air, is to "detect, identify, intercept, and destroy threat aircraft penetrating friendly airspace" (Hinton, 1994).

The F-22 design is driven by the aircraft's principal mission, which is offensive counter air (TAF, 1991).[1] Advanced technologies, including stealth,[2] are intended to enable the F-22 to penetrate deeply into hostile airspace and shoot down threatening aircraft before they can detect the F-22. The F-22 has been designed to acquire the enemy target and shoot first before being acquired. The design philosophy has also emphasized that, if acquired, the F-22 would have

---

[1] A member of the committee who is very familiar with Air Force operations has pointed out that a discrepancy appears to exist between various definitions of the term "offensive counter air." The definition in the Basic Aerospace Doctrine of the United States Air Force (Air Force Manual 1-1) can be read to include the destruction of enemy aircraft on the ground. The definition given in briefings to the committee by representatives of the Air Combat Command and the F-22 System Program Office (Hinton, 1994 and Raggio, 1994) and a discussion in the F-22 Operational Requirements Document (TAF, 1991) state the F-22 mission only in terms of engagements with enemy aircraft in the air (i.e., with no defined air-to-surface requirement). For the purpose of this report, the committee uses the definition that involves only air-to-air engagements.

[2] Stealth characteristics are intended to make it difficult or impossible for enemy sensors to acquire or bring weapons to bear on the system.

INTRODUCTION

countermeasures, speed, and maneuverability to minimize the likelihood of being hit.

> *First look, first shot, first kill. . . .F-22 [is] designed to evade hostile fire.*
> —Brigadier General William Hinton, Director of Requirements, Air Combat Command

The Committee on the Study of Live Fire Survivability Testing of the F-22 Aircraft was charged to study the desirability of waiving, for the F-22, the survivability tests required by law (10 U.S.C. 2366). As part of the study, the committee was asked to determine the sufficiency of the current vulnerability test program for the F-22. The committee was not asked to evaluate the total survivability program for the F-22 and did not do so. However, to assess the vulnerability program, it is important to examine vulnerability within the context of overall survivability.

The survivability[3] of a weapon system in combat is determined by several factors. First is its ability to avoid detection and interception in performing its mission. If a weapon system is detected and attacked, a second factor is its ability to use attributes like maneuverability to avoid being hit. Factors like these determine *susceptibility*. Finally, if a weapon system is hit, the inability to complete its mission, return to base safely, or ensure the safety of its personnel is called *vulnerability*. The survivability of a weapon system is increased by decreasing susceptibility, vulnerability, or both.

The F-22 program has placed primary emphasis on decreasing susceptibility, albeit with a significant vulnerability reduction program that includes design and testing. This emphasis has been driven by the advent of stealth technology, advances in avionics, and the blending of other new technologies. Also, the offensive counter-air mission of the F-22 tends to allow avoidance of denser surface-to-air threats at lower altitudes.

When the characteristics of a system do not prevent its acquisition and attack by the enemy, the second aspect of survivability—vulnerability—becomes more significant. The Department of Defense determines how far system designers and manufacturers should go to incorporate vulnerability reduction techniques. Some of the ways system vulnerability can be reduced are through redundancy (e.g., redundant structural load paths and electronics), fuel tanks that cannot explode when hit, fire suppression systems, use of nonflammable fluids, and armor (i.e., shielding of vital components and crew).

---

[3] An acceptable definition of survivability, which was used in the previous National Research Council report *Vulnerability Assessment of Aircraft* (NRC, 1993), is as follows: "The capability of an aircraft to avoid and/or withstand a man-made hostile environment (Ball, 1985)."

INTRODUCTION

The subject of this report is F-22 vulnerability, not susceptibility. However, both are significant aspects of survivability, and neither should be considered in a vacuum by decision makers. Both aspects influenced the committee's recommendations, even though susceptibility was not explicitly evaluated.

Use of the F-22 for roles other than offensive or defensive counter air, such as surface-attack operations, is being considered as a future possibility. Other missions and new threats that affect aircraft configuration, performance, and radar signature (for example) could make the terms of the survivability equation extremely dynamic—influencing the balance between susceptibility and vulnerability—and require expeditious and prudent consideration of vulnerability design and testing. Major modifications resulting from any new mission would require the Air Force to reassess vulnerability and implementation of the law on live fire testing.

VULNERABILITY TESTING OF AIRCRAFT VERSUS GROUND VEHICLES

The law governing live fire testing had its origin in ground vehicle vulnerability. One should be careful, however, to differentiate between a fighter that has a mission of offensive counter air—with its high dependency on stealth, countermeasures, speed, and maneuverability—and a ground vehicle that is difficult to make stealthy and is not fast or agile. The relative weights given to susceptibility and vulnerability and their realization in a design are different for fighters and ground vehicles (or even for aircraft with other missions). Also, it is more difficult to simulate, accurately and completely, the environment experienced by a fighter than that experienced by a ground vehicle.

With respect to susceptibility and vulnerability, high-performance fighters are the result of a finely tuned optimization—in which aerodynamic shape, lifting capacity, payload volume, structural weight, avionics, and hydraulic systems are put into a balance along with manufacturing costs, reliability, maintenance, and repair—that is significantly different from ground vehicles. The effect of any vulnerability deficiency corrections on susceptibility (and, for that matter, on such elements of system performance as payload, range, and speed) must be considered. Changes in many aspects to reduce vulnerability are considerably more likely to affect system performance adversely for aircraft than for ground vehicles.

In simulating the operational environment, two important considerations are (1) the stress state of the aircraft structure, normally a great deal higher than that of ground vehicles, and (2) the potentially large aerodynamic forces that can result when surface materials are distorted into the airstream or holes of significant dimensions expose inner airframe spaces to free airstream pressures. Both considerations imply significantly more complex and expensive live fire testing for aircraft, if all potentially important phenomena are to be represented fully in such tests (e.g., the imposition of realistic loads and sufficiently high air velocities before arranging projectile impact).

REPORT ORGANIZATION

The report is organized to reflect the subtasks in the Statement of Task (set forth in the Preface). Chapter 2 includes a discussion of circumstances affecting the F-22 survivability program that have occurred since Milestone II, which is the decision point that initiated engineering and manufacturing development of the F-22. Chapter 3 deals with questions concerning practicality and affordability of full-up, full-scale live fire testing. Chapter 4 addresses the sufficiency of the F-22 test program. Chapter 5 treats vulnerability assessment tools, which are very important to survivability testing of the F-22 as well as other aircraft. Finally, Chapter 6 presents the committee's recommendations.

There are, in addition, four appendices. Appendix A provides information about the meetings, site visits, and discussions of the committee. Appendix B is the Live Fire Test Law (10 U.S.C. 2366). Appendix C provides the text of documents associated with the Department of Defense F-22 waiver request. Appendix D reproduces pertinent material from the National Research Council's 1993 report *Vulnerability Assessment of Aircraft*.

REFERENCES

Ball, R.E. 1985. (Cited in NRC, 1993.) The Fundamentals of Aircraft Combat Survivability Analysis and Design. New York: American Institute of Aeronautics and Astronautics, Inc.

Hinton, W.S. 1994. Threat, Mission, and Operational Requirements for the F-22. Presentation to the Committee on the Study of Live Fire Survivability Testing of the F-22 Aircraft, National Academy of Sciences, Washington, D.C., December 21.

NRC (National Research Council). 1993. Vulnerability Assessment of Aircraft: A Review of the Department of Defense Live Fire Test and Evaluation Program. Air Force Studies Board, NRC. Washington, D.C.: National Academy Press.

Raggio, R.F. 1994. Overview of the F-22 Program. Presentation to the Committee on the Study of Live Fire Survivability Testing of the F-22 Aircraft, National Academy of Sciences, Washington, D.C., December 21.

TAF (Tactical Air Force). 1991. Advanced Tactical Fighter (ATF) System Operational Requirements Document (SORD) TAF 304-83-I/II-A. March 1. (S/NF)

2

Origin of Testing Requirements

This chapter discusses the legal requirements for live fire testing, the interpretations of those requirements within the Department of Defense (DoD), and the circumstances that led to the request for a retroactive waiver for the F-22. The chapter also addresses F-22 design requirements related to vulnerability. Finally, the committee offers some observations about the disposition of recommendations made by the previous National Research Council (NRC) committee that studied vulnerability assessment of aircraft (NRC, 1993).

FRAMEWORK

The ultimate objective of any combat system is to destroy or disable the enemy before being destroyed or disabled by that enemy. In meeting that objective, one consideration is for the combat system to have some level of inherent battle damage tolerance. Thus, aircraft designers must give serious attention to reducing vulnerability as part of the overall aircraft design.

Once a system enters the formal acquisition process, company designers receive contractual specifications from the system program office of the designated service. The system program office (the developer) bases its specifications on system requirements defined by the user.

In establishing F-22 survivability requirements, the system Operational Requirements Document (ORD) was the operative source (TAF, 1991). The F-22 ORD reflects the Systems Threat Assessment Report (STAR) (FASTC, 1992), which defines the threat the F-22 is expected to face based on its planned mission. The user's emphasis in the ORD was on susceptibility rather than vulnerability.

Experience in the 1980s with the Army's Bradley Fighting Vehicle led many to conclude that survivability of equipment and personnel had not been adequately considered and tested. As a consequence, in fiscal year 1987 Congress amended Title 10 of the U.S. Code by adding the original live fire test (LFT) law (10 U.S.C. 2366, 1986), which mandated certain procedures in vulnerability testing. Authority that permits the Secretary of Defense to waive the application of the tests under certain defined circumstances was included in the law. (The current version of the LFT law is provided in Appendix B.)

An overview of the guidelines, opinions, and amendments pertinent to the original law appears below. This overview sets the context for considering the F-22 waiver currently requested by the DoD.

F-22 LIVE FIRE TESTING REQUIREMENTS

This section addresses the many documents that influence the F-22 live fire test program. Because the early documents were reviewed in detail by the previous NRC committee (NRC, 1993), much of the information below is taken from that committee's report.

The Live Fire Test Law—Requirements and Historical Interpretations

The following historical information addresses the law's intent:

> The intent of the LFT law is to determine the inherent strengths and weaknesses of adversary, U.S., and allied weapon systems sufficiently early in the program to allow any design deficiency to be corrected. According to the FY1988-1989 DoD Authorization Act Conference Report, Congress intended that the Secretary of Defense implement the LFT law "in a manner which encourages the conduct of full-up vulnerability and lethality tests under realistic combat conditions, first at the sub-scale level as they are developed, and later at the full-scale level mandated in the legislation" (U.S. Congress, 1988).
> (NRC, 1993)

For purposes of this report, the present committee adopts the previous NRC committee's definitions of full-up and full-scale tests:

- A full-up test is defined as "a test conducted on a complete system or a partial system, with the full complement of fuel, ammunition, and hydraulic fluid carried by the system into combat."
- A full-scale or complete system test is defined as "a test conducted on the complete or total system, with or without the full complement of fuel, ammunition, and hydraulic fluid carried into combat."

The previous committee left no doubt about the requirements of the LFT law:

Based upon the evidence gathered by the committee and its study of the law, the committee is unanimous in the opinion that the LFT law requires a full-scale, full-up aircraft to be tested, regardless of the outcome of the sub-scale tests, unless a waiver is granted.

This committee agrees with the above opinion. In retrospect, it would appear that the LFT law should have been clear to everyone with little chance for misinterpretation. However, in the discussion that follows, it becomes obvious that serious misinterpretations *did* occur, which were not fully corrected until approximately 1993.

Without going into the nuances of past interpretations and misinterpretations of the law resulting from guidance issued by the DoD, it is instructive to review the prior NRC committee's statements:

[The committee] believes that the definitions in the 1988 Guidelines and the guidance given in the 1989 Planning Guide are not sufficiently clear as to the law's requirement that full-scale, full-up testing must be conducted. As a consequence of this misunderstanding, the Services have proceeded with sub-scale Live Fire Test programs on several weapon systems without making a provision for testing a complete system and without asking for a waiver because of the belief that no full-scale, full-up testing was required if early tests on sub-scale targets showed no design weaknesses.

The F-22 LFT program, among others, was covered in briefings to the prior NRC committee. Live fire testing of a full-up, full-scale F-22 was not planned (NRC, 1993).

The F-22 passed Milestone II in mid-1991 and entered engineering and manufacturing development. No waiver from the law's requirements was granted by the Secretary of Defense before the F-22 entered this phase. Yet there is evidence to suggest that some in the Air Force knew, in 1991, that a waiver was required.

In October 1991, Brigadier General James A. Fain, who was then Director of the F-22 System Program Office (SPO), sent a memorandum to the Office of the Secretary of the Air Force outlining F-22 live fire test alternatives. With respect to full-up testing of the system configured for combat, General Fain recommended "requesting a waiver to this requirement" (Fain, 1991; see below).

In August 1992, James O'Bryon, the Deputy Director of Test and Evaluation for Live Fire Testing in the Office of the Secretary of Defense, sent a memorandum to the Air Force Director of Test and Evaluation. With respect to the planned F-22 Live Fire Test and Evaluation Vulnerability Program, Mr. O'Bryon stated:

Although in many respects [the ballistic] part of the proposed plan appears to be comprehensive and well thought out, in one respect at least it falls seriously short of the requirements of the law. . .. In June of last year, I held a series of meetings with then BG Fain, Director of the F-22 SPO. I left those meetings with the clear understanding that full-up testing would be included in the Live Fire Test Program . . .. I am concerned that, after more than a year, this most important inadequacy, the lack of full-up testing, has not been addressed.
(O'Bryon, 1992)

The committee notes that Fain's memorandum to the Office of the Secretary of the Air Force preceded O'Bryon's memorandum by nearly a year, indicating that O'Bryon was unaware of Fain's waiver recommendation. The SPO's position is discussed further in a later section of this chapter.

With respect to waivers, the previous NRC committee (NRC, 1993) recommended that:

. . . the Director, Test and Evaluation, formalize the waiver process by developing a risk-benefit assessment methodology that can be used uniformly to determine whether a full-scale, full-up test program for any particular aircraft is "unreasonably expensive and impractical." The methodology must also be applicable to the evaluation of the alternate Live Fire Test program for the sub-scale targets.

. . . the Secretary of Defense take measures to ensure (a) that the LFT&E [Live Fire Test & Evaluation] Guidelines are properly enforced by requiring either that covered systems be subjected to full-scale, full-up testing or that a waiver be obtained; (b) that any waiver be fully justified; (c) that the waiver process be uniformly applied; and (d) that no stigma be attached to the use of the waiver process.

Recent Live Fire Test Guidelines and Interpretations

A new set of DoD guidelines, issued January 27, 1994, superseded all previous editions. Revised definitions of full-up test and full-up system-level test are presented in those guidelines, as follows (Longuemare, 1994):

Full-up Test. A vulnerability test conducted on a complete or partial system loaded or equipped with all dangerous materials (including flammables and explosives) that would normally be on board in combat (configured for combat). All critical subsystems which could contribute to the test outcome must be operating (e.g., hydraulic and electrical power) under realistic conditions. . .. This testing alone may not satisfy 10 USC, Section 2366. . . .

Full-up, System-Level Test. A LFT&E Strategy for a covered system, major munition program, or missile program, or covered product improvement program will include Full-up, System-level tests. The term "Full-up, System-level Test" is that testing that fully satisfies the statutory requirement for "realistic survivability testing" or "realistic lethality testing" as defined in Section 2366, Title 10, USC.

On July 25, 1994, the Air Force published Instruction 99-105 to provide "guidance and procedures for the live fire test and evaluation of Air Force systems" (USAF, 1994). The document states that it is "in compliance with LFT&E legislation." It also states:

LFT&E is a sub-set of developmental test and evaluation (DT&E); the Air Force accomplishes it through a balanced program of test and analysis. Analysis must be an integral part of the LFT&E process because it is unreasonably expensive and impractical to test all possible combinations of threats, aircraft configurations, and environments. The Air Force must initiate the LFT&E program sufficiently early to allow the results to impact system design prior to full-rate production or major modification. The primary emphasis is on testing and analysis to identify and correct potential design flaws. The LFT&E program should provide an assessment of a system's vulnerability or lethality characteristics relative to the expected spectrum of battlefield threats.

Regarding LFT&E waivers, the instruction states:

The Secretary of Defense may waive LFT&E legislation requirements for covered systems, major munitions programs, and product improvements (that significantly affect vulnerability or lethality) to covered systems and major munitions programs where LFT&E would be unreasonably expensive and impractical. Request, process, and have a granted waiver before Milestone II. The Air Force cannot grant LFT&E waivers after Milestone II, except through legislative relief.

An approved LFT&E waiver exempts a system, program, or product improvement from full-up, system-level testing (full scale, complete system configured for combat, equipped with all fluids, materials, and explosives). However, you must have an alternate plan to evaluate system vulnerability or lethality (such as combinations of analysis, component, sub-system, or sub-assembly testing, etc.) to fulfill the intent of the LFT&E legislation.

The committee believes that, with the exception of the requirement for full-up, full-scale testing, the requirements for waived LFT&E programs are no less stringent than for nonwaived programs.

Recent Amendment to Waiver Provision of the Live Fire Test Law

The waiver provision of the LFT law was amended in 1994 by addition of the following (10 U.S.C. 2366; see Appendix B for full text of the waiver provision):

> [T]he Secretary may waive the application of the survivability and lethality tests of this section to such a system or program and instead allow testing of the system or program in combat by firing munitions likely to be encountered in combat at components, subsystems, and subassemblies, together with performing design analyses, modeling and simulation, and analysis of combat data. . . .

In making this change, the congressional conference committee stated that the amendment (Congressional Record, 1994):

> . . . would make it clear that the certification which must be provided to Congress . . . must be submitted before the system enters engineering and manufacturing development. The effect would be to maintain realistic survivability and lethality testing through testing of components, subsystems, and subassemblies in cases where the Secretary waives requirements for full up testing . . . .

REQUEST FOR A RETROACTIVE F-22 TEST WAIVER

The committee's third formal task (see Preface for Statement of Task) was to discuss what happened after Milestone II to cause the F-22 program manager to request a waiver. The short answer is an improved understanding within DoD of the legal requirement.

Milestone II for the F-22 program occurred in June 1991 (Raggio, 1994); the provision of the law that permits a waiver dates to 1987; and a waiver for the

F-22 program was requested in 1993. The waiver request is discussed below, followed by the committee's assessment of what happened.

The DoD General Counsel submitted draft legislation to Congress on October 8, 1993, "[t]o authorize a retroactive waiver of the survivability and lethality testing procedures that apply to the F-22 program" (Gorelick, 1993a, 1993b). The draft legislation was accompanied by a plan for alternative assessment of the vulnerability of the F-22 aircraft. The General Counsel's letter (as addressed to the President of the Senate) and its attachments are reproduced in Appendix C of this report. Briefly, DoD justifies its proposal as follows (Gorelick, 1993a, 1993b):

> Because of the cost of an F-22 aircraft, such [live fire] testing is both unreasonably expensive and impractical. Since the F-22 has already entered full-scale engineering development, legislation is needed to allow the Secretary of Defense to grant a waiver. . . .. [T]he Air Force developed the revised live fire test program that . . . includes detailed analyses, review of historical test data, and incremental build-up testing that includes material characterization tests and live fire testing of selected components and subassemblies. Information from the results of these tests will be taken into account in the F-22's design.

The SPO identified two areas for potential aircraft loss: (1) fire/explosion within the dry bay and/or fuel tank ullage (i.e., the volume of the tank above the fuel) areas, and (2) hydraulic-ram-induced structural failure of the fuel cells (see Chapter 4). The Air Force cites the conduct of limited ballistic tests, directed energy tests, and vulnerability model enhancements as demonstrating its efforts in vulnerability reduction methods.

There has been no technical change in the program to account for the change in the Air Force's position. The history of what happened is best understood in terms of four points:

- the Air Force position on vulnerability;
- the Air Force position on full-up, full-scale live fire testing;
- confusion over the interpretation of the law; and
- hope within DoD and the Air Force that the previous NRC committee would help mitigate the requirements of the law.

Each of these points is discussed below.

Position on Vulnerability

Air Force briefings to this committee expressed the view that F-22 design emphasis should be placed on reducing susceptibility as opposed to reducing vulnerability (Hinton, 1994; Leaf et al., 1994). Briefers also expressed the opinion that system weight was of overriding importance and should not be increased significantly to achieve vulnerability reductions.

The Air Force position, at least that voiced to the committee, is that (a) F-22 qualities like speed, maneuverability, and ability to kill the enemy first are more important to survivability than vulnerability (Hinton, 1994); and (b) further major improvements in vulnerability reduction would likely be won only by more serious losses in these other qualities (Ogg, 1995).

Position on Full-Up, Full-Scale Testing

To the committee's knowledge, the Air Force has never planned to conduct full-up, full-scale tests of the F-22, believing that they would be too expensive and result in the loss of aircraft needed for other test purposes. The Air Force estimates that full-up, full-scale live fire testing of the F-22 would cost approximately $250 million, an amount that has not been budgeted (Raggio, 1994). Chapter 3 discusses this matter in more detail.

Confusion over Interpretation of the Law

Confusion over DoD interpretations of the law has already been discussed. During the previous NRC committee's deliberations in 1991 and 1992, various service representatives and contractors expressed their understanding that then current guidelines indicated that full-up, full-scale, live fire testing was not really required (NRC, 1993). However, someone in the legal office of the Office of the Secretary of Defense (OSD) had other ideas, since the SPO (i.e., General Fain, see above) wrote to the Office of the Secretary of the Air Force in October 1991 (Fain, 1991):

> We have been informed that OSD's legal council [sic] determined that full-up "system configured for combat" testing . . . is required to meet the intent of Chapter 139 of Title 10, United States Code, Section 2366, major systems and munitions programs: survivability and lethality testing; operational testing. We recommend requesting a waiver to this requirement.

The events surrounding this correspondence from General Fain pose something of a puzzle:

- According to James O'Bryon (O'Bryon, 1992), he and General Fain met in June 1991, near the time of Milestone II. Mr. O'Bryon then had the clear understanding that full-up testing would be included in the F-22 live fire test program.
- According to Charles "Pete" Adolph, General Fain never mentioned live fire testing when he briefed the Defense Acquisition Board for Milestone II in June 1991.[1]
- General Fain's correspondence seems to indicate a clear appreciation by the SPO in October 1991 of the requirements of the law.
- Finally, while General Fain's correspondence in October of 1991 was prepared after Milestone II in June 1991, a long time elapsed between that recommendation to request a waiver and the waiver activity in 1993.

At least part of the explanation for this sequence of events seems related to expectations of help from the previous NRC committee.

Help from the Previous Committee

Between mid-1991 and early 1993, when the previous NRC committee was conducting its study, the Air Force and some DoD people apparently hoped that the previous committee would help mitigate the requirements of the law by recommending to Congress that the requirement for full-up, full-scale, live fire testing be reduced (Hawley, 1994).[2] Publication of the NRC's January 1993 report, which upheld the need for a waiver, was a critical mining point.

Several Air Force briefings to this committee referred to the recommendations of the previous NRC committee and the subsequent DoD guidelines (Hawley, 1994; Ogg, 1995; and Raggio, 1994). Others, who were OSD officials in 1991, expressed disappointment that the previous NRC committee did not take issue with the need for full-up, full-scale, live fire testing of aircraft (Reed, 1995):

---

[1] As indicated in a conversation on February 8, 1995, among Mike Clarke and Harry Reed of the Committee on the Study of Live Fire Survivability Testing for the F-22 Aircraft and Charles "Pete" Adolph, Larry Stanford, and Al Rainis, all of whom were OSD officials involved in test and evaluation at the time of Milestone II.

[2] The February 8, 1995, conversation with Mr. Adolph substantiated this point. See note 1.

They [Adolph, Stanford, and Rainis] were of the general opinion that the legislation is not appropriate for most of the Air Force's aircraft, which are not expected to be highly able to survive close encounters with bullets and warheads. Although they all felt that the legislation had added some value to the design process for aircraft, they expressed disappointment that the NRC LFT Committee had not challenged portions of the law that bear on the amount of knowledge acquired versus the cost of full-up testing as opposed to what was learned through the component and sub-assembly testing.

A Lingering Question

In spite of all of these circumstances, there is the lingering question of why the SPO did not request a waiver before Milestone II (when the Secretary of Defense could have granted it) to be on the safe side. The answer to this question may lie in the services' reluctance to submit waiver requests, which the previous NRC committee detected, leading it to recommend "that no stigma be attached to the use of the waiver process" (NRC, 1993).

F-22 DESIGN REQUIREMENTS FOR VULNERABILITY

Design requirements that relate to F-22 vulnerability are based on its counter-air missions—principally, the conduct of offensive counter-air operations (see Chapter 1). The offensive counter-air mission placed a premium on achieving high survivability through reduced aircraft susceptibility. Reduction of vulnerability, although important, was deemed less significant.

There is, however, an explicit statement in the 1994 Test and Evaluation Master Plan that indicates one air-to-surface mission has already been added.[3] Even with this mission, the F-22 is to use internal carriage of weapons and fly at

---

[3] The plan (TEMP, 1994) states:
The F-22 SPO has been directed to integrate the 1000 pound Joint Direct Attack Munition (JDAM) on the aircraft in support of the F-22's Air to Ground capability. The F-22 SPO is currently working with ACC [Air Combat Command] and the F-22 contractor team in further defining the actual program plan. When this planning is complete and placed on contract, program plans which will include test planning requirements will be available and incorporated into future releases of the TEMP [Test and Evaluation Master Plan]. Therefore, the January 1994 TEMP includes no reference to the JDAM integration.

a relatively high altitude. Nonetheless, the LFT law requires weapon systems with major modifications to undergo appropriate live fire testing or request a waiver. Presumably, a significant change in the F-22's mission (e.g., to an air-to-surface mission, which could increase the aircraft's exposure to surface-to-air defenses) would be considered a major modification that would necessitate additional vulnerability assessments.

The ORD (Operational Requirements Document) identifies two general vulnerability requirements (TAF, 1991). The first is that flame-retardant hydraulic fluid and fuel-tank inerting be employed. The second is that avionics, power, and hydraulics be redundant, fault tolerant, and damage tolerant. In addition, quantitative design requirements with respect to vulnerability are specified in the contract for the F-22. These contract specifications are needed to optimize aircraft design by addressing various tradeoffs and to assess contractor performance (Stewart and Shipman, 1995).

The F-22 SPO did an engineering assessment of the state-of-the-art of vulnerability reduction technology and the desired attributes of the aircraft. The SPO then specified vulnerability design requirements in two ways. First, general design requirements (e.g., minimize fuel dumping into the engine inlet ducts) were specified. Second, the SPO specified that the total vulnerable area of the F-22 should not exceed certain levels, expressed in square feet, against various threats.[4] The resulting vulnerable areas were based on analyses of the F-22 design and were accepted since they compared favorably to the vulnerable areas of the F-15 against the same threats (Stewart and Shipman, 1995). These vulnerable areas were coordinated with and accepted by the user and established as the contractual design values.

The values in these specifications *do not include the effects of on-board ordnance*. This assumption is addressed again in Chapter 4.

Specifications for vulnerability to high-energy microwave devices (stated in frequencies and maximum power levels) are also included in the F-22 contract (Giorlando, 1995). There are no specifications for protection against laser devices other than for the F-22's missile systems and the pilot's vision. The SPO considers that pilot safety is satisfied by the current visor.

In summary, the user chose to put more emphasis on reducing susceptibility, the probability of being hit in the first place, than on reducing vulnerability, the probability of being killed if hit. The users' choices were reflected in the specifications developed by the SPO. For future F-22 missions (e.g., air-to-surface), the Air Force should reexamine the balance of requirements among susceptibility, vulnerability, and related performance parameters.

---

[4] Vulnerable areas are defined as the product of the probability of kill given a random hit (by a fragment or a bullet) and the presented area of the target aircraft (NRC, 1993).

DISPOSITION OF PRIOR COMMITTEE'S RECOMMENDATIONS

The recommendations of the previous NRC committee fall within three categories (NRC, 1993): (1) clarifying and enforcing the waiver process; (2) improving the vulnerability data base and methodology, to include the development of cost-benefit analyses; and (3) organizational and administrative changes in the way OSD approaches the LFT&E program and vulnerability assessment in general.

This committee is pleased that OSD has in fact clarified the waiver process. The committee regrets that a cost-benefit analysis process has not been developed, although the Air Force is making some progress in this regard (Griffis, 1995). The committee also strongly advocates that several areas of vulnerability technology receive considerably more emphasis: the development of an empirical data base, development of models, and experimental validation of these models. Finally, the committee does not wish to take up the organizational and administrative issues—partly because that would take the committee far from its stated mission and partly because some of the reorganizations within OSD would make the recommendations moot.

SUMMARY

Requirements Background

The events that led to the late (post-Milestone II) request for a waiver of the F-22 from full-up, full-scale live fire tests are complex. The original LFT law was passed in fiscal year 1987; since then there has been inconsistent interpretation of what live fire testing included. The most stringent interpretation was that a fully configured (with armament, fuel, etc.) combat system must be used. A less strict interpretation was that various sub-scale tests and models could be used to simulate the full-up, full-scale test and thereby provide information on system vulnerability. This committee (in agreement with the previous NRC committee) believes that the most stringent interpretation of the law is correct.

The currently accepted interpretation, as contained in the 1994 DoD guidelines, is quite clear on the requirements for full-up, full-scale, live fire testing or for a waiver from such testing prior to Milestone II. These requirements were subsequently reflected clearly in Air Force Instruction 99-105. If a waiver is granted, the law instead permits live fire testing of components, subsystems, and subassemblies to maintain realistic survivability testing.

What Changed to Cause Request for Waiver

Nothing specific to the aircraft program appears to have changed since Milestone II in 1991 that would have prompted the Air Force to request a waiver. However, there appears to be a changed view of what is required to fulfill the intent of the LFT law as it is interpreted within the DoD. The committee's interpretation of the events is as follows.

To the committee's knowledge, the Air Force never planned to conduct full-up, full-scale, live fire testing on the F-22. Prior to the NRC report in 1993, the Air Force believed that the component and subassembly testing protocol met the intent of the law. The Air Force initially based its belief on the DoD guidelines and then apparently hoped that the previous NRC committee would help mitigate the full-up, full-scale requirement (allowing instead the component and subassembly test methodology). Although there was some consideration of a request for a waiver in 1991, it was not until after the NRC report in 1993 that the Air Force appeared to accept fully the fact that a waiver was necessary. By then Milestone H had long since passed (in mid-1991) for the F-22.

F-22 Design Requirements for Vulnerability

The F-22 was engineered principally on the basis of its offensive counter-air mission, which emphasized reduced susceptibility over vulnerability reductions. The user did not specify quantitative vulnerability requirements. Instead, contract specifications for vulnerability were established by the SPO and accepted by the user.

The existing vulnerability specifications and the live fire test program do not address future missions (e.g., air-to-surface) that the F-22 might be required to perform. For future F-22 missions, the Air Force should reexamine the balance of requirements among susceptibility, vulnerability, and related performance parameters.

Disposition of Previous Recommendations

The committee is pleased that the waiver process has been clarified but regrets the lack of more progress in developing a cost-benefit methodology. Vulnerability technology still needs considerably more emphasis on the development of data bases and the development and validation of models.

REFERENCES

Congressional Record. 1994. 103d Cong., 2d Sess., 140(120): H8932.

Fain, J.A. 1991. Correspondence to Office of the Secretary of the Air Force, Acquisition, October 21. F-22 System Program Office.

FASTC (Foreign Aerospace Science and Technology Center). 1992. F-22 System Threat Assessment Report (STAR), November 23. (S).

Giorlando, J. 1995. High Power Microwave. Presentation to the Committee on the Study of Live Fire Survivability Testing of the F-22 Aircraft, F-22 System Program Office, Wright-Patterson Air Force Base, Ohio, January 19.

Gorelick, J.S. 1993a. Letter to the Honorable Thomas Foley, Speaker of the House of Representatives, October 8. Washington, D.C.: Office of the General Counsel of the U.S. Department of Defense.

Gorelick, J.S. 1993b. Letter to the Honorable Al Gore, President of the Senate, October 8. Washington, D.C.: Office of the General Counsel of the U.S. Department of Defense.

Griffis, H. 1995. Cost Benefit Analysis Methodology. Presentation to the Committee on the Study of Live Fire Survivability Testing of the F-22 Aircraft, Dayton, Ohio, January 20.

Hawley, R.E. 1994. Discussion of Waiver of Live Fire Tests. Presentation to the Committee on the Study of Live Fire Survivability Testing of the F-22 Aircraft, National Academy of Sciences, Washington, D.C., December 21.

Hinton, W.S. 1994. Threat, Mission, and Operational Requirements for the F-22. Presentation to the Committee on the Study of Live Fire Survivability Testing of the F-22 Aircraft, National Academy of Sciences, Washington, D.C., December 21.

Leaf, H.W., R. Lauzze, and J. Ogg. 1994. Air Force Test and Evaluation Plans for F-22 and Other Applicable Air Force Aircraft. Presentation to the Committee on the Study of Live Fire Survivability Testing of the F-22 Aircraft, National Academy of Sciences, Washington, D.C., December 22.

Longuemare, R.N. 1994. Live Fire Test and Evaluation Guidelines. Washington, D.C.: U.S. Department of Defense, Acquisition and Technology.

NRC (National Research Council). 1993. Vulnerability Assessment of Aircraft: A Review of the Department of Defense Live Fire Test and Evaluation Program. Air Force Studies Board, NRC. Washington, D.C.: National Academy Press.

O'Bryon, J. 1992. Memorandum to Director, Air Force Test and Evaluation, August 17. F-22 Live Fire Test and Evaluation (LFT&E) Vulnerability Program. Washington, D.C.: Office of the Secretary of Defense.

Ogg, J. 1995. Vulnerability Program Overview. Presentation to the Committee on the Study of Live Fire Survivability Testing of the F-22 Aircraft, at the F-22 System Program Office, Wright-Patterson Air Force Base, Ohio, January 19.

Raggio, R.F. 1994. Overview of the F-22 Program. Presentation to the Committee on the Study of Live Fire Survivability Testing of the F-22 Aircraft, National Academy of Sciences, Washington, D.C., December 21.

Reed, H. 1995. Personal correspondence to Mike Clarke, Al Rainis, and Larry Stanford, February 15, on a discussion that took place February 8, 1995, at Science Applications International Corporation, Tysons Comer, Va., between Reed, Clarke, Adolph, Stanford, and Rainis.

Stewart, M., and J. Shipman. 1995. Ballistic Vulnerability Analysis. Presentation to the Committee on the Study of Live Fire Survivability Testing of the F-22 Aircraft, F-22 System Program Office, Wright-Patterson Air Force Base, Ohio, January 19.

TAF (Tactical Air Force). 1991. Advanced Tactical Fighter (ATF) System Operational Requirements Document (SORD): TAF 304-83-I/II-A. March 1. (S/NF).

TEMP. 1994. F-22 Test and Evaluation Master Plan. Wright-Patterson Air Force Base, Ohio : F-22 System Program Office. January.

USAF (U.S. Air Force). 1994. Interpretation of Live Fire Test Law. Air Force Instruction 99-105. Washington, D.C.

U.S. Congress. 1988. (Cited in NRC, 1993.) FY 88-89 DoD Authorization Act Conference Report, Live-Fire Testing (Section 802).

3

Practicality, Affordability, and Cost-Benefit

This chapter presents the committee's views on the practicality of full-up, full-scale testing for the F-22. It also discusses affordability and cost-benefit aspects of conducting and not conducting these tests. Finally, the committee's opinion on the desirability of the F-22 waiver is expressed in the context of its conclusions on practicality, affordability, and cost-benefit.

PRACTICALITY

The committee defines an activity as "practical" if meaningful results can be achieved with existing technology and reasonably available resources. The issue facing the committee is whether it is practical to conduct full-up, full-scale, live fire tests of the F-22. Several considerations were taken into account in addressing this issue; they serve to organize the discussion of practicality that follows.

Relative Importance of Vulnerability Reduction to F-22 Survivability

The live fire test law's definition of "realistic survivability testing" states that "the primary emphasis [is] on testing vulnerability with respect to potential user casualties and taking into equal consideration the susceptibility to attack and combat performance of the system" (10 U.S.C. 2366). Judgments about the practicality of full-up, full-scale tests for the F-22 must therefore take into account its susceptibility and combat performance.

The probability of any system surviving hostile action is usually expressed as the difference between unity (a perfect return rate if hostile action is ineffective) and the product of the probabilities representing the effectiveness of hostile action. In other words, the probability of survival is defined as $1-P_D(P_{H/D})(P_{K/H})$, where P_D is the probability of being detected; $P_{H/D}$ is the probability of being hit by the enemy's weapon, once detected; and $P_{K/H}$ is the probability of the friendly system being rendered permanently and completely ineffective, once hit. It is clear that the lower the value of any one of these three probabilities—P_D, $P_{H/D}$, or $P_{K/H}$—with the other two held constant, the greater the friendly system's chance of survival.

The last of these probabilities, $P_{K/H}$, is the metric influenced by reducing vulnerability and dealt with in planning and executing live fire tests.

As indicated earlier in this report, the F-22 design has been optimized for its primary mission of offensive counter air. The aircraft includes both offensive and defensive capabilities that are expected to result in destruction of enemy aircraft beyond visual range—before the enemy aircraft can locate the F-22 and fire a missile. Only a very small proportion of encounters are expected to result in closure to ranges at which enemy missiles or guns could be fired at the F-22. The F-22 has been designed with a high degree of maneuverability to deal with the occasional close engagement.

Accordingly, F-22 survivability depends more on low values of susceptibility (i.e., the product of P_D and $P_{H/D}$) and less on low values of vulnerability ($P_{K/H}$). Any option to reduce F-22 vulnerability further must be evaluated for its effect on combat performance (e.g., low susceptibility), with priority generally going to preserving the performance.[1]

Of course, the priority on low susceptibility does not rule out design changes to address a critical vulnerability discovered in the live fire test program and attendant analysis and other testing. The issue is not whether a live fire test program is required—all agree that it is. The issue is how far to go with the live fire test program. For the F-22, this judgment must be influenced in part by the relatively low weight the Air Force has given to vulnerability in the overall survivability equation.

Chapter 2 cited indications that the F-22 will also have an air-to-surface mission in the future. As currently envisioned, that mission will involve delivery of munitions from a relatively high altitude. An air-to-surface mission could affect the terms in the survivability equation. Specifically, the vulnerability term might assume greater significance because of increased exposure of the F-22 to surface-to-air defenses. Future missions for the F-22 will require that the Air Force reassess the relative importance of vulnerability and susceptibility and adjust the vulnerability assessment program accordingly.

---

[1] High-performance fighters result from a highly integrated and carefully balanced optimization of a substantial variety of systems. Since survivability is influenced by all three terms (P_D, $P_{H/D}$, and $P_{K/H}$), the aircraft designer should not make changes to any one of them without considering its effects on the other two. For example, if it were possible to reduce $P_{K/H}$ by armor plating an avionics bay, but in so doing the armor increased the aircraft's radar reflectivity, raising its P_D, and if the weight of the armor significantly reduced the aircraft's maneuverability, raising its $P_{H/D}$, then the improvement in $P_{K/H}$ could be ill advised. This kind of one-dimensional improvement activity involves the dangers of suboptimization.

No consideration of vulnerability reduction can be definitive without being specific about the threat, which can influence all three terms in the survivability equation. The next section discusses threat realism, and Chapter 4 describes the threat environment projected for the F-22 and how that environment is reflected in the vulnerability assessment program.

Realism in Aircraft Testing

The committee believes that live fire testing can be conducted at four levels of aircraft readiness for the intended mission. These levels, which reflect various degrees of realism, are discussed below. *Even the highest level does not adequately simulate actual flight conditions.*

Level 1.

The first and lowest level is live fire testing of hardware simulations or mock-ups of production systems. For example, the hydraulic elements of a flight control system might be assembled in the proper full-scale geometry and provided with flight-condition operating pressures, but mounted on a working representation of an airframe. The committee believes that such tests are encompassed by the waiver provision of the live fire test law [see Appendix B, Section 2366(c)(2)].

Level 2.

The second level involves full-scale aircraft components, subsystems, or subassemblies representative of production items. Depending on the purpose of the test, all hardware that would be present in the production configuration may or may not be present in the component, subsystem, or subassembly that is tested. The committee considers these to be the tests referred to in the waiver provision of the live fire test law [Section 2366 (c)(2)].

Level 3.

The third level of live fire testing involves a complete, production aircraft not loaded with live ordnance or fuel. Selected systems may or may not be operating. The committee considers this to be inert (not full-up), full-scale testing. A waiver would be needed if only this type of full-scale system testing were planned.

Level 4.

The fourth and highest level utilizes a complete production aircraft with all systems, fuel, and live ordnance installed and in operation typical of combat. A full-up, full-scale test can be considered a verification (based on a random sample) of the results of the lower level tests. Tests on the aircraft would be conducted in ground-test facilities where test conditions can be closely controlled and flight conditions can be approximately simulated. With currently available facilities, it is impossible to generate the air flows and actual design stress levels encountered in flight. Parts of the aircraft can be bathed in high-speed subsonic air, and some lower states of stress can be simulated. Tests under these conditions would, in the opinion of the committee, come the closest to meeting practically the congressional intent for full-up, full-scale testing.

It is desirable, of course, to make any test as realistic as reasonable. In live fire tests of aircraft, realism can be thought of as having three independent aspects, as discussed below: the threat, the configuration of the friendly system, and the operational status of the friendly system. If cost were not a consideration, complete realism, or something close to it, might be achieved in all three aspects. But since costs are in fact a determinative constraint, it is important to consider the cost of approaching realism in any one aspect in the light of realism achievable in the other two.

> Subassemblies
> With respect to the sizes or types of subassemblies that could be used, the committee notes that the LFT law does not specify any particulars in connection with its use of the word "subassemblies." Subassemblies could be a wing, a wing box, a complete or partial fuselage, etc. The important point is that the subassemblies be of sufficient size and composition to represent adequately the vulnerability effects for which the test is being conducted. At various places in this report the committee uses terms like "large assemblies," the "major parts" of an aircraft, "assemblies from early production," and "major subassemblies." These expressions merely reflect the committee's view that the subassemblies used should be large enough and sufficiently representative of a production configuration to meet the live fire test objectives.

Threat realism involves a wide range of variables (e.g., if the threat is explosive, variables include the distance from its target upon detonation, the kinematics of the intercept, and the characteristics of the blast generated). Because testing the virtually infinite number of possibilities is obviously out of the question, definition of the threat involves the assignment of probabilities to all of these characteristics of weapon-target relationships. The indeterminable features of threat definition are at the root of the statistical nature of the analysis needed to interpret live fire test results. In complex ways, such considerations must enter into the determination of $P_{H/D}$ and influence measured or computed values of $P_{K/H}$.

Achieving realism *in the aircraft's configuration* is straightforward, but it can be expensive. The possibility of discovering an unexpected interaction between systems argues that everything to be carried on a mission be in place during live fire testing. This is the full-up aspect, which includes fuel, ammunition, and hydraulic fluid. In addition to correct size, shape, and hardness, the full-scale aspect requires that all components be installed (e.g., wire bundles and batteries). Although the aircraft's configuration when hit by hostile fire cannot be predicted, it will usually be possible to predict the most vulnerable condition (e.g., live ordnance and fuel). There is little that cannot be determined about the friendly aircraft's configuration for live fire testing.

Achieving realism *in the aircraft's operating conditions*, however, is another matter. Some operational circumstances can be replicated in live fire tests at reasonable cost, but others cannot. For example, hydraulic systems can be pressurized and engines can be running in ground tests. On the other hand, the combined effects of high-speed airstreams and maneuvers may make the results of ground tests a poor indicator of what will happen under dynamic flight conditions (e.g., a high-g turning pull-up).

The variability introduced by indeterminable aspects of the threat and operating conditions leads to one of the most important considerations in achieving realism, namely, the concept of statistical significance. Statistical significance quantifies the confidence that one can place in the test results obtained. This confidence is a function of the variability of the data and the size of the sample set, which is the number of trials that can be conducted in the case of live fire testing.[2] It is important to be able to support adequate sample sizes to achieve reasonable statistical significance across all levels of testing. Vulnerability assessment models and simulations provide a way of achieving relatively large sample sizes at relatively low cost. However, current models and simulations have their own sets of problems, which are discussed later (see Chapter 5).

Taking all of the above into account, the committee believes there is no feasible way known to test the vulnerability of a first-line fighter like the F-22 in a fully realistic way, that is, as defined by law. It might be possible to conduct a one-or-few-trials flight test with drone aircraft. But the variability associated with threats and operational conditions would require tens of trials for meaningful results. Although large numbers of trials in the full-up, full-scale configuration could conceivably be feasible for obsolete fighters, they are not practical for an expensive new operational system such as the F-22. The difficulty (in some cases, inability) of conducting realistic tests must be considered when planning live fire test programs for the F-22 and evaluating their results.

---

[2] It is important to recognize, in particular, that the results of a single test do not provide statistically significant information about low probability events, such as "unknown unknowns." These events are not likely to occur in a single test (or even a few tests). For example, if an event has a 10 percent probability of occurrence, and one wants 95 percent confidence that it will be observed in the test sequence, then the number of test trials needed is about 28 (NRC, 1993). With half the number of tests, the confidence of observing the event drops to 75 percent. With one-fourth the number of tests (i.e., 7), the confidence drops to 50 percent.

Destructive Versus Nondestructive Testing

Destructive testing[3] can be of great value in any developmental effort. Such testing can be conducted at any of the levels discussed previously. However, by itself, destructive testing of a complete aircraft at the full-scale level yields extremely low confidence factors for probabilistic outcomes due to the small number of trials possible. At lower levels of testing (e.g., the level that involves components), greater numbers of destructive trials become possible. On the other hand, some aspects of realism (e.g., the synergistic effects associated with a full-up, full-scale test program) may be lost at these levels. Even a full-up, full-scale destructive test would not in itself provide meaningfully representative information.

Nondestructive tests can also be conducted at all levels with varying degrees of realism. These tests can provide a sufficient number of trials for reasonable confidence at reasonable costs. For example, nondestructive and repeatable full-scale tests with nonflammable, nontoxic materials may in certain circumstances provide useful data.

The committee believes that there is a major difference between destructive and nondestructive testing of the F-22. When nondestructive tests can be designed (e.g., with high-power microwaves[4]) it appears feasible to test the full-scale aircraft in sufficient trials to make a meaningful assessment of vulnerability. The committee supports such testing.

The committee's support does not extend to full-up testing of the F-22 in situations that might detonate any live ordnance or fuel on board. Provisions to bypass the destructive features would, of course, make the tests less than full-up.

Expert Opinion

Finally, the opinion of most experts with whom the committee discussed the matter, and the committee's opinion, is that full-scale testing of a complete aircraft is much less likely to provide useful information than are appropriate component, subsystem, and subassembly tests. This opinion is based on three factors:

---

[3] Destructive testing is commonly understood to be the opposite of nondestructive testing, which is an approach to testing that does not involve damage or destruction of the test sample. By its very nature, live fire testing causes damage to the material or component being tested. In fact, the extent of damage is one of the key results of a live fire test. The committee adopts the common understanding discussed here in its use of the terms "destructive" and "nondestructive" testing.

[4] The committee notes that high-power microwaves can cause some damage to components, so even these kinds of tests may not be considered truly nondestructive.

PRACTICALITY, AFFORDABILITY, AND COST-BENEFIT

1. A full-scale test with a threat weapon sufficiently large to affect the whole aircraft may not be repeatable over a reasonable range of conditions unless many aircraft samples are used. The committee judges that the benefits of such an approach would not be worth the costs for an aircraft like the F-22.
2. If a munition with a local effect is used for the test, then a component, subsystem, or subassembly can be used as the test specimen. A buildup test sequence could be used that (a) starts at the component level, where the greatest number of trials are possible; and (b) moves to the subsystem and then the subassembly or large assembly level, as appropriate. This approach would permit sufficient trials for developing and confirming (or disproving) various hypotheses about what damage might or might not occur at the different levels. The collective wisdom (including hard data) that is established by this process provides reasonable confidence in the results.
3. There is a need to understand damage and failure mechanisms at the component, subsystem, and subassembly levels.

The committee was briefed by at least one individual (O'Bryon, 1994) who stated his strong belief that testing at the full-scale level is necessary, using production-authentic hardware and systems, but without live ordnance aboard. He is correct that the current F-22 live fire program does not go this far.

Specific arguments raised in the above individual's presentation for accomplishing full-scale (if not full-up) testing of the F-22 were essentially (a) there are secondary effects (ricochet, debris, spalling, etc.) that do not reveal themselves in smaller scale tests; (b) synergistic effects [where damage to one subsystem may cause damage elsewhere, sometimes called cascading effects (see Chapter 4)] should be determined by full-scale tests; (c) system degradation should be measured as the result of such tests; (d) battle-damage repair insights should be encouraged; (e) fire starting mechanisms should be observed, and fire suppression should be evaluated; and (f) "unknown unknowns" can occur during full-scale tests. These arguments appeared to represent not only this individual's views but also the views of others who advocate full-up, full-scale testing.

The committee carefully weighed the array of arguments on both sides of the full-scale testing issue. The committee was persuaded that, for a system like the F-22, considering its mission and system characteristics, a step-by-step approach to vulnerability assessment was best. A test plan that dictates a methodical buildup of tests from the component level, to the subsystem level, to the subassembly level, to the large assembly level, and, if required, to the full-scale level made more sense to the committee than assuming at the outset that full-scale tests were necessary in every case. Additionally, while each of the enumerated benefits of full-scale testing is important, in the opinion of the committee and others who briefed the committee, most can be achieved without resorting to full-scale testing.

The philosophical issue involved is that tests must proceed to a level that, when taken collectively with the results of all the lower level tests, produces a reasonable likelihood of revealing all necessary information. To some, that dictates full-scale testing; others disagree. It is true that full-scale tests have a potential for disclosing surprises that no one could predict through modeling, analysis, or component-to-subassembly testing. However, for the F-22, especially in light of its very high cost, a test without a reasonable expectation of additional valuable data to be derived is unwarranted. The committee did not discern a reasonable expectation of deriving valuable data (e.g., "unknown unknowns") from a full-scale test of the F-22. If it had, the committee would have had no qualms about recommending such testing.

The committee concluded, in light of its own expert judgment and the prevailing opinion of most others with whom the committee met, that full-scale testing of an aircraft like the F-22 is not justified. However, the committee does agree that testing of F-22 subassemblies in a complete or nearly complete production configuration is justified in appropriate circumstances, as discussed in Chapter 4.

AFFORDABILITY

The committee defines an "affordable" activity as one within budgetary constraints or attainable budgets. Like all acquisition programs, the prioritization of alternatives within the F-22 program is based on an assessment of their marginal utility. Unaffordable budget items remain unfunded until either more funding is authorized or the priorities of activities within the program are reassessed (e.g., previously unaffordable items displace funded items based on analyses that reassess their relative benefits).

The committee was not able to consider fully or challenge the prioritization of items within the F-22 program budget. However, based on the committee's judgment of benefits versus costs, even if an aircraft could be provided, the full-up, full-scale tests would not be recommended. As a result, the overall program budget and prioritization of expenditures within that budget, as established by the F-22 SPO, was accepted by the committee. Clearly, in the SPO director's estimation, full-up, full-scale testing exceeds the available budget (Raggio, 1994).

Affordability of Full-Up, Full-Scale Testing

According to data provided by the SPO as of February 1995, the cost of the currently planned F-22 live fire tests is slightly over $38 million (then-year dollars). This amount funds multiple component, subsystem, and subassembly tests. The SPO projects that a full-up, full-scale test of the F-22 would cost an additional $250 million (then-year dollars) above the currently planned program. The major component of this amount is purchase of a full-up production aircraft.[5] The underlying assumption is that this test asset would be devoted fully to the tests and could not reasonably be refurbished to have military utility after completion of the tests.

The committee requested information to support the contention that a production aircraft would be required in lieu of some less realistic alternative with somewhat less test fidelity (e.g., the prototype test aircraft). The SPO responded that use of three alternatives had been examined: (1) the prototype air vehicle, (2) the static ground test article, and (3) an engineering and manufacturing development flight test aircraft (Graves, 1995). None of these was deemed feasible in the judgment of the SPO. The SPO's estimate that the cost of a full-up, full-scale test would be on the order of $250 million results directly from this judgment. The committee has no basis for refuting the SPO's judgment.

There is an argument that $250 million is only a small percentage (less than 0.5 percent) of total F-22 program costs, and therefore cost should not be a determining factor. The committee understands this view. Nonetheless, the committee's judgment is that the benefits of full-up, full-scale tests are not commensurate with the costs. Even if $250 million were provided for additional vulnerability assessment of the F-22, the committee would not support using the funds for full-up, full-scale testing.

Both Air Force and Navy experts on aircraft vulnerability assessment indicated that, if they were provided a new production aircraft for live fire testing, they would prefer to disassemble it and perform live fire testing on a less than full-up, full-scale configuration. They stated that their preference to test at the component, subsystem, and subassembly levels was derived from their experience that they would learn more about vulnerability, and could place greater confidence in the results, for the resources expended.[6]

---

[5] The full-up aircraft represents well over 90 percent of the $250 million cost estimate (Graves, 1995).

[6] The opinions in this paragraph were expressed during discussions with members of the committee on February 21, 1995, at the Naval Air Warfare Center, China Lake, California.

Investment Methodology for F-22 Vulnerability Tests

The committee was briefed by the SPO on an investment methodology that examined incremental live fire testing of the F-22 (Griffis and Lauzze, 1995). The committee was not persuaded by the SPO's analysis. The committee believes that the SPO's investment model offers a limited context from which to reject, analytically, additional increments of live fire testing. Analytic approaches based on incorrect premises have a way of producing logically derived results that may well be wrong. Additional comments on cost-benefit methodology appear below.

COST-BENEFIT METHODOLOGY

The committee received briefings on the current state of the art of cost-benefit analysis of live fire testing (Griffis and Lauzze, 1995; Klopcic, 1995). Attempts are under way within DoD to develop a methodology for determining the return on investment of successive levels of live fire tests by predicting the costs of future aircraft attrition. If the cost of an additional live fire test is known (it is), and one is reasonably sure of the incremental reduced cost of attrition that a test would bring about (one is not), the return on investment would be clear and the additional test could be judged on that basis.

The methodologies presented are immature at this time. They appear to address suboptimal measures of benefit. Despite these shortcomings, the committee believes that a validated, useful methodology for determining, quantitatively, the cost-benefit relationships of live fire testing would be a valuable tool for vulnerability assessment. This kind of tool would help prioritize the various levels of live fire testing together with other competing program activities.

The committee acknowledges that validating a useful cost-benefit framework is, in fact, extremely difficult because of the need to establish a priori benefit valuations. There is a danger that a framework could produce misleading results if the benefit measures chosen are so narrow as to preclude interesting alternatives. The committee suggests that this risk be minimized by conducting a series of excursions that assess proposed test programs in the light of alternative measures as well as widely varying but conceivable test results. In addition, it might be useful to consider military scenarios that would elevate the importance of reduced vulnerability[7] and, thereby, possibly enhance the benefit-to-cost ratio of given levels of testing.

---

[7] Here, the concept of reduced vulnerability could extend to fewer losses of flight crews (i.e., capture or death) and smaller likelihood that U.S. systems would be recovered and exploited by the enemy.

Additional live fire testing of the F-22 at any level should be examined in terms of opportunities for information to be gained, test costs, and conceivable consequences of not performing the tests. Such testing should not be rejected because of cost-benefit scores in a single, and perhaps overly simple, construct of how the future may develop.

For example, assume that (a) the United States will be faced with a series of wars like Desert Storm extending through the first half of the twenty-first century; and (b) the F-22 will confront significantly improved defensive systems, will be heavily used, and will be at risk in each sortie it flies. In the face of these assumptions, the total number of combat sorties flown is the key measure. Reduced vulnerability and increased capability to repair battle-damaged aircraft[8] may become determinative. Under these conditions, additional investment in live fire testing could be attractive, particularly if it might reveal ways to produce a significant increase in the number of combat sorties flown (e.g., through increased tolerance to damage sustained in combat and the development of techniques for repairing damaged aircraft in the field).

CONCLUSIONS

Having weighed the arguments on both sides of the full-scale testing issue, the committee was persuaded that, for a system like the F-22, a well conceived, incremental build-up of tests that proceed from the component level to the subassembly or large assembly levels made the most sense. The committee did not discern a reasonable expectation of deriving additional valuable data (e.g., "unknown unknowns") from a full-scale test of the F-22.

The committee concludes that completely realistic, destructive, full-up, full-scale testing of the F-22 is not practical and entails high costs relative to the resulting benefits. The judgment of most members of the live fire test community with whom the committee met is that incremental testing to the point of relatively large subassemblies is the way to proceed and that full-up, full-scale tests of combat-configured aircraft are of marginal utility. The committee agrees with this judgment. The combination of the lack of realism in test conditions, the difficulty of obtaining a sufficient number of trials, and expert opinion all support this conclusion.

Full-up, full-scale testing in a configuration that could destroy the entire aircraft if a detonation occurs is therefore not warranted for the F-22. On the other hand, nondestructive testing (e.g, with high-power microwaves) is practical, as is destructive testing of components, subsystems, and subassemblies. The committee

---

[8] The F-22's new composite materials and new systems could require battle-damage-repair techniques that are much different from those used on current fighters.

endorses the use of production-representative articles for these tests. Also, it is practical to test parts of an aircraft under simulated loads and exposure to high-velocity airflows.

Regarding the affordability of the tests, the committee has concluded that affordability is not the matter of foremost relevance. Cost-benefit is most relevant. Affordability only becomes relevant if the benefits relative to the costs of whatever tests are being considered are commensurate with the benefits relative to the costs of other alternatives. With respect to full-up, full-scale tests for the F-22, the committee judges the benefits to not be worth the costs.

Based on its conclusions concerning the impracticality and low benefits for the costs of full-up, full-scale, live fire testing for the F-22, the committee is unanimous in its opinion that a waiver is the appropriate course of action for the F-22. It must be pointed out, however, that while the committee was asked to examine practicality, affordability, and cost-benefit, the law states that a waiver may be granted by the Secretary of Defense based on a certification "that live-fire testing . . . would be unreasonably expensive *and* impractical [emphasis added]." The committee's interpretation is that both conditions must be true before a waiver can be granted. Regarding the term "unreasonably expensive," the committee believes that the low benefits relative to costs (i.e., high costs relative to benefits) means that the tests are unreasonably expensive.

If a waiver is granted, there needs to be some measure of sufficiency for the test program that is conducted. The sufficiency of live fire tests currently planned for the F-22 is addressed in the next chapter.

Finally, the committee recognizes that its judgments regarding the costs and benefits of full-up, full-scale testing were reached in the absence of a mature methodology for assessing benefits relative to costs. The committee is leery of reliance on cost-benefit methodologies that use an overly simple construct of the F-22's future to make judgments about how far to go with live fire testing. A broader analytical framework could elevate the importance of reduced F-22 vulnerability over the long haul and might enhance the benefit-to-cost comparisons of given levels of testing.

REFERENCES

Graves, J.T. 1995. National Research Council Questions on Live Fire Test. Memorandum from Deputy Director, F-22 System Program Office, to National Research Council, Mike Clarke, February 14.

Griffis, H., and R. Lauzze. 1995. Cost Benefit Analysis Methodology. Presentation to the Committee on the Study of Live Fire Survivability Testing of the F-22 Aircraft, Dayton, Ohio, January 20.

Klopcic, J.T. 1995. Knowledge-Based Benefit/Cost Methodology for Live Fire Test Evaluation. Presentation to the Committee on the Study of Live Fire Survivability Testing of the F-22 Aircraft, National Academy of Sciences, Washington, D.C., February 16.

NRC (National Research Council). 1993. Vulnerability Assessment of Aircraft: A Review of the Department of Defense Live Fire Test and Evaluation Program. Air Force Studies Board, NRC. Washington, D.C.: National Academy Press.

O'Bryon, J.F. 1994. Discussion of Live Fire Testing Philosophy and the History Associated with First Report. Presentation to the Committee on the Study of Live Fire Survivability Testing of the F-22 Aircraft, National Academy of Sciences, Washington, D.C., December 21.

Raggio, R.F. 1994. Overview of the F-22 Program. Presentation by to the Committee on the Study of Live Fire Survivability Testing of the F-22 Aircraft , National Academy of Sciences, Washington, D.C., December 21.

4

Sufficiency of F-22 Testing Plans

Part of the committee's task was to evaluate the sufficiency of the F-22 test program to meet the requirements of the live fire test law. This chapter contains that evaluation.

Discussed first are the F-22 threat environment and its replication by the SPO in the vulnerability assessment program. The vulnerability assessment program is then evaluated. Finally, after some additional observations, the committee presents its conclusions.

F-22 THREAT ENVIRONMENT AND ITS REPLICATION

The threat environment for the F-22 is derived from its current principal mission of conducting offensive counter-air operations. The environment consists of threats the aircraft would expect to face while accomplishing its mission of destroying enemy aircraft over hostile territory. These threats, as characterized by a representative of the Air Combat Command in an unclassified briefing to the committee (Hinton, 1994), are shown in Table 4-1. In addition, the vulnerability specifications (discussed in Chapter 2) reflect a high-power microwave threat and a laser threat.

The committee accepts these threats for the counter-air missions. The aircraft has been designed to meet them, and its vulnerability assessment program has been structured accordingly.

The following assumptions were made by the SPO in the live fire test program to replicate the threat:

- The threat spectrum for the F-22 is represented by: (a) two metallic fragments (45 grains and 150 grains), (b) two armor-piercing incendiary (API) rounds (23mm and 30mm), and (c) two high-explosive incendiary (HEI) rounds (23mm and 30mm) (SPO, 1995a). The effects of the fragments are to be assessed over a range of impact velocities from 2,000 to 9,000 feet per second (fps). The API rounds will be assessed over a range of impact velocities from

TABLE 4-1 F-22 Threat Environment

Fighters	Air-to-Air Missiles	Surface-to-Air Missiles
Current		
Mirage 2000	Reticle IR Seeker with CCM	SA-10
Gripen	AA-10B/D	SA-12
MiG-29 Fulcrum	AIM-9M	
SU-27 Flanker	AA-7D	
	Active-Radar Seeker	
	AA-X-12	
	AIM-120	
	Multi-element Seeker	
	AA-11	
	PYTHON 4	
	MAGIC 2	
Future		
New Fighters IOC by 2004	Imaging IR Seeker AAM	SA-X-17
SU-35 Improved	(IOCs in 2004)	(notional)
Flanker	XAAM-4	
Rafale	ASRAAM	
Eurofighter 2000	AIM-9X	
New (Notional) Fighters		
Available by 2014		
Multi-role Fighter		
Interceptor IOC 2005-2008		
Experimental Fighter Interceptor		
IOC 2010-2015		

Source: Hinton, 1994.
NOTE: AAM=air-to-air missile; CCM=counter-countermeasures; IOC=initial operational capability; IR=infrared.

500 to 4,000 fps, and the HEI rounds will be assessed for a velocity of 2,500 fps.
- For each threat that has a range of velocities, the maximum vulnerable area (over the range of velocities) is calculated for each of the 6 cardinal views (top, bottom, front, back, left, and right). In the case of the HEI rounds, the vulnerable areas are calculated for each cardinal view for the single velocity.
- These maximum vulnerable areas are averaged over the 6 cardinal views to provide the maximum-allowable vulnerable area for each threat. Specifically, 36 numbers (6 for each of the 2 fragments plus 6 for each of the 4 cannon rounds) are averaged to produce 6 numbers (i.e., vulnerable areas) that are incorporated in the F-22 contract as vulnerability specifications. Vulnerabilities for specific encounters with specific weapons can be calculated using the same methodology, but they are not part of any formal requirement.

The committee considered the SPO's assumptions. While the threat missiles may well change, the fragments, with the spread in velocity, seem to be a robust representation of the effects of individual fragments from missile warheads. The API and HEI rounds are reasonable representations of air-to-air cannon-fired rounds.

The sole use of vulnerable areas for individual fragments to represent missile warheads does imply certain assumptions and limitations:

- The fragment data are only intermediate data. The effect of a particular warhead event is a much more complicated affair involving details of the warhead and the end-game geometry (i.e., guidance and control capabilities of the missile and the action and signature of the target).
- The effects of multiple fragment hits are ignored (consistent with the current state-of-the-art of vulnerability analysis).
- The usual assumption is made that blast effects are only important for miss distances at which the fragments would certainly kill the target. However, there are intercept geometries for which this assumption is not the case, and blast must be taken into account in the analyses.

When anti-air missiles detonate, a spray of fragments, often focused in a specific direction, is propelled from the warhead to the target. In addition, a significant blast wave caused by the detonation of the warhead also propagates toward the aircraft. The fragments and blast wave will strike the aircraft at different times, sometimes resulting in enhanced kill mechanisms.

The committee believes that the two discrete fragment sizes selected by the SPO for analysis and test are representative of the fragments from the spectrum of warheads likely to be encountered. However, vulnerability of the F-22 to these fragments is estimated on a one-fragment-at-a-time basis and therefore, in itself, does not completely represent the vulnerability of the aircraft to missile warheads. Vulnerability of the aircraft against these warheads is assessed in a subsequent analysis that considers three types of kill mechanisms:

- Blast kill of the structure based on overpressures resulting from detonation of the warhead's high-explosive charge.
- Impact of multiple fragments with the fragments spaced far enough apart so that their effects on the aircraft are independent; the result effectively is aggregated from independent single fragment assessments.
- Impact of multiple fragments that are dense enough so that their effects are not independent and, when taken together, could result in structural kill of the aircraft. This type of kill mechanism is often accounted for by a kinetic energy threshold for structural kill; it is particularly important for annular blast fragmentation or focused blast fragmentation warheads.

The first two of these mechanisms are accounted for in the vulnerability analysis currently being conducted for the SPO. The third has not yet been explicitly accounted for. It is safe to ignore this effect for many warheads and encounter geometries because, if such a kill is obtained, a kill from one of the other two mechanisms would also be obtained. However, for some classes of warheads (e.g., the annular or focused blast fragmentation warheads) this kill mechanism may be important and should be considered in future analysis and testing by the SPO.

OVERVIEW OF THE AIR FORCE VULNERABILITY ASSESSMENT PROGRAM

The committee received extensive briefings on the vulnerability assessment program from representatives of the F-22 SPO during its visit to Wright-Patterson Air Force Base in January 1995. Those briefings and communications with the SPO provided the basic information evaluated in this chapter.[1]

The SPO and its contractors performed a detailed vulnerability analysis of the F-22 using revised versions of standardized computer models (see Chapter 5). The outputs of the analysis were estimated vulnerable areas and overall system

---

[1] The entire briefing is documented here as SPO, 1995a. Several parts of this briefing are also cited individually in this chapter (e.g., Griffis, 1995a, and Ogg, 1995).

$P_{K/H}$ (see Chapter 3). The SPO then assessed the analytical results and identified areas of uncertainty. These areas of uncertainty were based on the following specific criteria:

- Areas currently treated as invulnerable based on analysis for which insufficient or contradicting data exist.
- Compartments where collateral damage mechanisms cannot be assessed and that represent a potential vulnerability.
- Components that represent a significant contribution to vulnerable area and have insufficient supporting data.
- Areas for which the basic material or ballistic data base is inadequate.

Areas of uncertainty were next mapped against the F-22 design and the results of the vulnerability assessment to identify test issues, areas of the aircraft that needed to be tested, and specific test hardware requirements. The enumerated F-22 live fire tests and locations of the test areas on the aircraft are shown in Figure 4-1.[2]

As a part of establishing its test program, the SPO made several assumptions (SPO, 1995a):

- Because the threat scenario is for offensive counter-air missions, 60 percent of total usable fuel is assumed to remain after penetration into hostile air space. According to the design fuel-burn sequence, the fuel in certain critical tanks would have been consumed. This assumption minimizes inlet fuel ingestion kills (see discussion later in this chapter) and reduces vulnerable areas by 30 percent.
- The flight is straight and level at 500 knots, and the requirements are independent of altitude.
- The kill category considered is the attrition kill, in which controlled flight is lost within five minutes following the hit.
- Vulnerable areas used in these calculations are based in part on results of individual tests against electronics modules. This approach may not take flammability of the coolant into account and may need revision pending the results of the tests on fluid flammability that are recommended later in this chapter.
- Concern for survivability of the pilot is demonstrated by a double barrier between the cockpit and the forward fuel tank. There are no

---

[2] Tests 9 and 10 are exceptions; Test 9 is a components-related test and Test 10 is a materials test. Neither of these tests appears in the diagram because they cannot be isolated to a specific location on the aircraft. Also, Test 5 and Test 8 do not appear in the SPO's numbering system because those tests were subsumed by Tests 4 and 7, respectively.

SUFFICIENCY OF F-22 TESTING PLANS 49

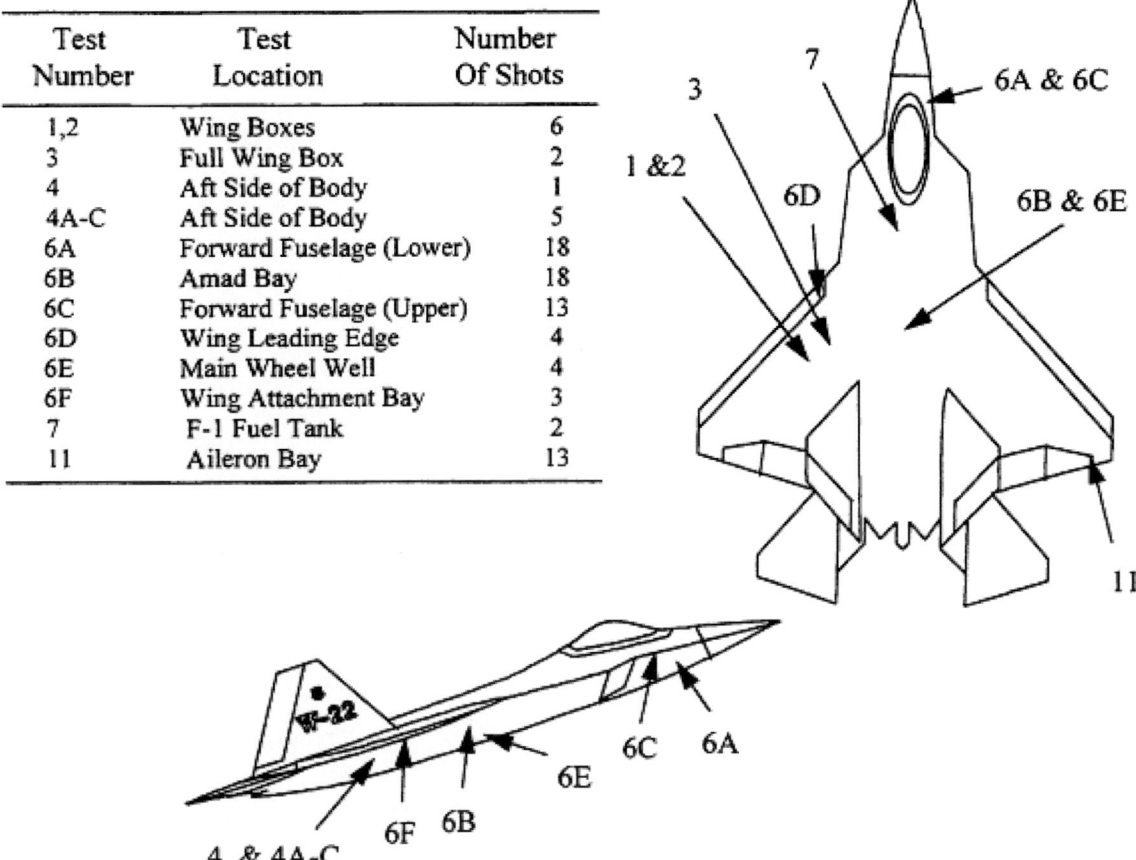

Test Number	Test Location	Number Of Shots
1,2	Wing Boxes	6
3	Full Wing Box	2
4	Aft Side of Body	1
4A-C	Aft Side of Body	5
6A	Forward Fuselage (Lower)	18
6B	Amad Bay	18
6C	Forward Fuselage (Upper)	13
6D	Wing Leading Edge	4
6E	Main Wheel Well	4
6F	Wing Attachment Bay	3
7	F-1 Fuel Tank	2
11	Aileron Bay	13

Figure 4-1 Locations of the test areas. Source: Griffis, 1995b.

detailed, formal specifications for pilot survivability, but there is a qualitative concern for maximizing the chance that the pilot can eject in case of catastrophe (Griffis, 1995a). The pilot is included in the assessment of vulnerable area.

- The specifications *do not include the effects of on-board munitions*. The reason given for this omission is that the design values are to be used to assess contractor performance, and the contractor is not responsible for the munitions (SPO, 1995a). However, mission analyses calculating $P_{K/H}$ of the aircraft account for the vulnerable area of on-board munitions.

The committee accepts several of these assumptions but has the following comments about the lack of specifications covering on-board ordnance. It is difficult to consider that the *system* is any less than the sum of the *aircraft and its ordnance*. The impact of aircraft design on protection of the ordnance and on possible mitigation of the effects of damage could certainly be important to the survivability of the pilot, even if it were not so important to the survivability of the aircraft. Saying that the ordnance is not the responsibility of the contractor trivializes an important issue.

The SPO identified additional areas of uncertainty for which further testing would be beneficial if funding were available (Graves, 1995). These include tests on polyalphaolefin (PAO) coolant fluid flammability, hydraulic fluid flammability, and fuel flammability. A second series of tests is proposed for the weapons bay, using a simulator of the weapons bay of representative size and materials. These tests would involve ballistic testing of actual AIM-9 and AIM-120 rocket motors with both protected and unprotected bays to determine the effectiveness of protecting the weapons bay against fires using ablative materials.

EVALUATION OF THE VULNERABILITY ASSESSMENT PROGRAM

This section evaluates the current test program. The organization is by major F-22 subsystem. Each subsection briefly describes a subsystem and attendant vulnerabilities, the analyses and tests already conducted or planned by the SPO for that subsystem, and the committee's overall assessment. Finally, revisions are suggested to improve the program where the committee believes them to be desirable.

Many live fire tests are planned or have been performed on various components, subsystems, and subassemblies of the F-22. Test articles range from component prototypes to engineering and manufacturing development (EMD) hardware. The threat kill mechanisms are those discussed above.

Major F-22 Subsystems
Structure and Integral Fuel Tanks
Fuel System and Associated Dry Bays
Flight Control and Auxiliary Systems
Weapons Bay and Ordnance
Engines
Flight Crew
Fire Protection Systems

Structure and Integral Fuel Tanks

Description and Attendant Vulnerabilities

Airframe outer skins are generally quite thin, highly stressed, and only slightly resistant to penetration from missile warhead fragments and projectiles. However, penetration of the skin and failure of an interior member such as a wing spar or a fuselage frame does not necessarily mean loss of the aircraft. This is particularly true for the F-22, since it is designed with multiple (redundant) load paths and utilizes structural materials that have a high fracture toughness. The design represents a significant improvement over many past fighter aircraft, which were largely single load path structures and often utilized high-strength aluminum and steel alloys with very low fracture toughness.

Figure 4-2 illustrates the structural configuration of the F-22, a largely multiple load path construction. For example, the F-22 has multiple wing spars; several wing carry-through fuselage frames and bulkheads; three fin attachment frames; numerous additional fuselage frames; and various stringers, stiffeners, ribs, and other miscellaneous members. The materials are shown in Figure 4-3.

A design goal was that the structure be able to sustain the damage from defined threats without loss of the aircraft. The multiple load path design was intended to achieve this goal. The uncertainty in predicting hydraulic ram effects[3] in the integral tanks and the need for developmental testing of some of these structures were recognized.

While the committee agrees that the F-22 structure is predominately of multiple load path design, there are some exceptions. Such exceptions include (a) the horizontal tails, where each tail is supported by a single pivot shaft made of a titanium alloy (see Figure 4-2); and (b) the farthest aft fuselage frame (i.e., Frame 6 in Figure 4-2), which carries a significant portion of the horizontal tail and vertical tail loads across the fuselage.

---

[3] Ball (1985) gives this explanation of hydraulic ram effects:

When a penetrator enters a compartment containing a fluid, a damage process called hydraulic or hydrodynamic ram is generated. Hydraulic ram can be divided into three phases: the early shock phase, the later drag phase, and the final cavity phase . . .

The hydraulic ram loading on all of the wet walls of the tank can cause large-scale tearing and petalling, with openings very much larger than those made by the actual penetrator. The hydraulic ram loading can also be transmitted through attached lines, causing failure at fittings or other discontinuities in the lines. . . .

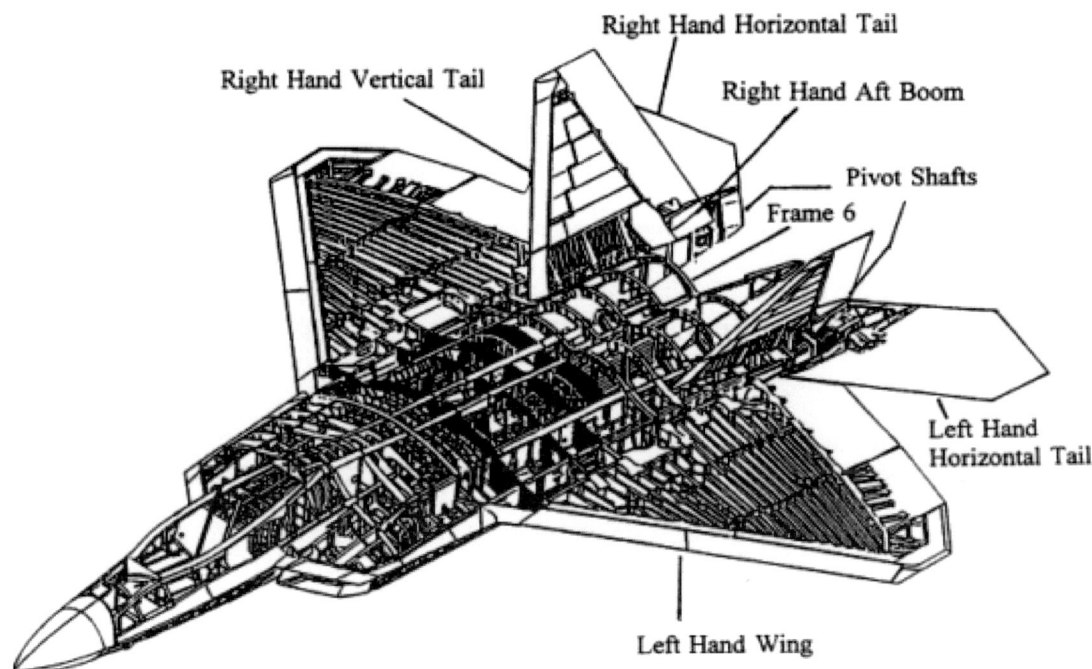

Figure 4-2 Structural configuration. Source: Griffis, 1995b.

Although the committee believes that the horizontal tail pivot shafts may be heavy enough to resist complete failure following ballistic impact, these shafts could encounter damage that would degrade their strength and ability to carry operational maneuvering loads. However, if failure should occur and a horizontal tail were lost, it is not certain that the aircraft would be lost. Although the F-22 prime contractor has indicated that the airplane can be controlled with one tail missing, it would appear that this capability would depend on flight conditions at the time of loss. Risk of aircraft loss given the loss of a horizontal tail has not been adequately defined.

While Frame 6 is not currently considered to be a vulnerable location, the committee believes that, if the frame fails as a result of a direct hit, loss of the aircraft might result. It appears that this potential vulnerability needs further investigation, as discussed below.

Also, the aft fuselage booms that support the horizontal tails as well as some of the vertical tail loads might be considered to be single load path even though

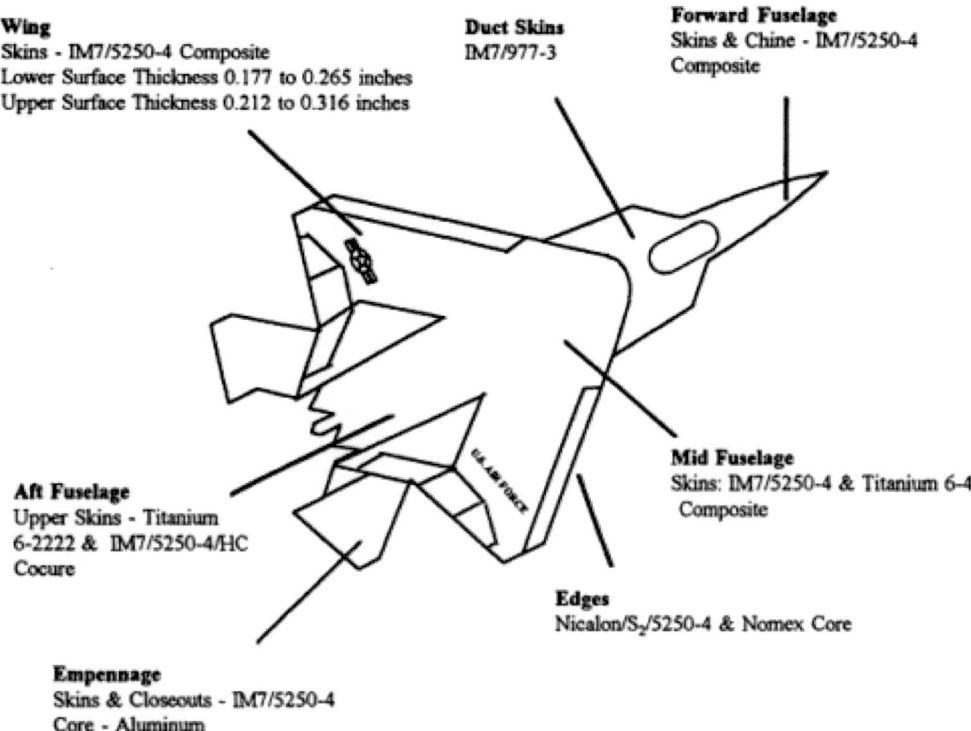

Figure 4-3 Materials applications. Source: Griffis, 1995b.

they consist of several structural members (see Figure 4-4). These members make up two beams, one on each side of the aircraft, that are subjected to substantial bending, shear, and torsional loads. If either of these boxes falls, it may lead to loss of the aircraft. As discussed below, some live fire testing has been done, and more is planned, for this general area.

When assessing the vulnerability of the structure to ballistic threats, it is necessary to consider the effects of other potential damage mechanisms in addition to projectile impacts. These include blast effects, overtemperature due to fire, and hydraulic ram loads in fuel-containing structures. Hydraulic ram can significantly magnify the damage to wing structure, and test results could lead to redesign. It is also the primary mechanism of concern in the fuselage fuel tank structures.

Of particular concern is the close proximity of the forward fuselage fuel tank (designated F-1) to the cockpit. The forward side and top of the tank has a double-walled barrier intended to prevent fuel leakage directly into the cockpit. The

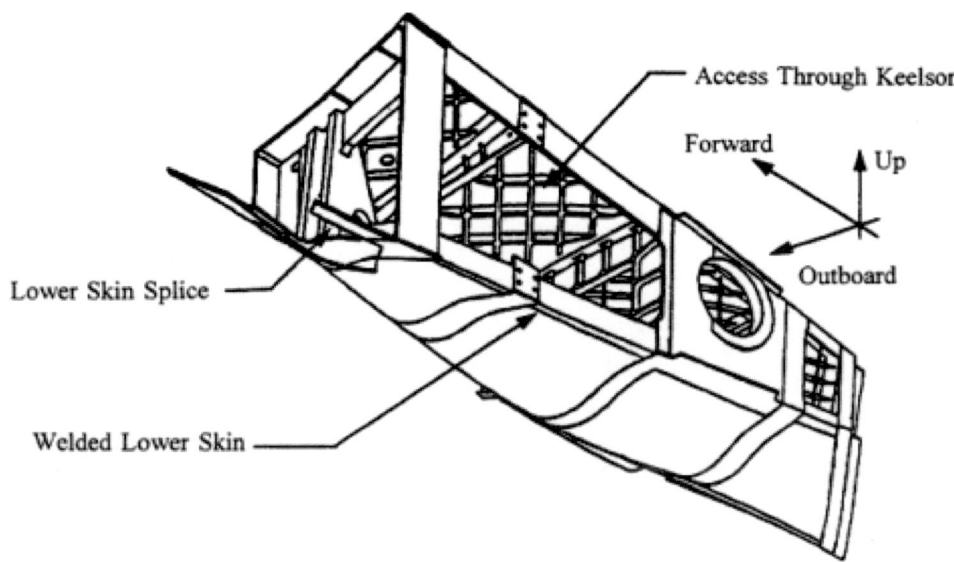

Figure 4-4 Aft boom. Source: Griffis, 1995b.

committee is concerned that, if the inner wall were ruptured as a result of hydraulic ram, the second barrier could be penetrated and fuel could leak into the cockpit. Fire could then break out because of the various ignition sources in the cockpit. Also, ram effects could damage the canopy hinge, actuator, support structure, seat rails, or other structures, which could then prevent successful ejection of the pilot.

Planned Analyses and Tests

The F-22 structure has been considered invulnerable because of the multiple load path design. However, there are some uncertainties in analytically predicting the extent of damage that will be encountered and, to a lesser extent, the residual strength of the remaining structure after it is damaged.

The uncertainty in predicting damage is probably greatest for fuel-containing structures (i.e., wings, aft sides of the body, and forward fuel tank) because of difficulties in predicting both the pressures and the response of the materials (particularly composite materials) associated with the hydraulic ram phenomena. It was because of this uncertainty that the developmental live fire tests of the wing structure were performed. Subsequently, hydraulic ram live fire tests were also planned for the aft fuselage and forward fuselage fuel-containing structures.

The original design damage size for the F-22 wing was estimated to be an 8-inch-diameter hole in the skin plus the loss of a spar. Accordingly, the wing was designed to tolerate this damage without loss of the aircraft. However, when the live fire tests were performed on a test box containing three composite spars (designated Test 1A), it was found that the damage was much more extensive (SPO, 1995a). In fact, the entire skin panel and all composite spars failed. This result led to redesign of the wing and the addition of five new titanium spars, as shown in Figure 4-5.

The redesigned configuration was then subjected to live fire testing using four-spar and eight-spar test boxes (designated Tests 1B, 2A, and 2C) (SPO, 1995a). In these tests, the damage was largely contained between the titanium spars (the composite skins and intermediate composite spars were both severely damaged) (Griffis, 1995a). Based on this result, it has been predicted that the wing can still carry the bending load associated with a 4-g maneuver (Ogg, 1995).

Verification of the final wing design is planned in a full-scale-wing live fire, residual-strength test program to be conducted between 1997 and 1998. This testing will be performed at the Air Force's Wright Laboratory, where the wing will be fueled and have simulated loads applied and air flowing over the wing during live fire testing. A residual strength test will then be performed on the damaged wing (SPO, 1995a).

The aft fuselage booms consist of a forward boom section, that contains fuel. It is supported by fuselage Frames 2 through 6 (see Figure 4-6) and the cantilevered section of the boom aft of Frame 6, shown in Figure 4-4, which is dry. These booms are fabricated from welded, integrally stiffened titanium (Griffis, 1995a).

The objective of Test 4 is to investigate structural damage to the aft fuselage forward boom area due to impact (SPO, 1993). Hydraulic ram effects (wet bays), blast effects (dry bays), and damage from fragments are investigated. All tests use 30mm HEI projectiles, since these are expected to generate the most severe pressures and fragmentation. As of this writing, three tests have been performed and one more has been planned.

A preliminary live fire test (Test 4A) was performed on a small welded box, which contained water, to determine if the impact would cause weld cracking. No cracking of the weld seams occurred, although the box was torn apart (SPO, 1993). This test was not considered to be a good hydraulic ram test because the box was made of nonrepresentative materials that failed during testing, thus relieving the pressure on the weld seams.

Test 4B was also performed in 1994 on a welded titanium box, which did not contain water or fuel. The purpose of this test was to obtain information on blast and fragment damage that could be expected if a 30mm HEI round

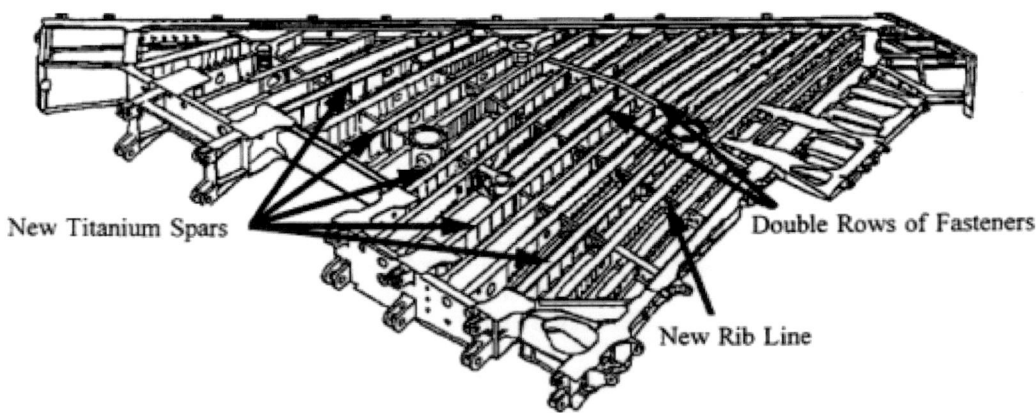

Figure 4-5 Current wing configuration. Source: Griffis, 1995b.

penetrated the aft dry boom (Griffis, 1995a). The results indicated that there was less fragment penetration than predicted by analysis. The damage did not appear to be severe enough to jeopardize flight safety should similar damage occur during an operational conflict. No further live fire tests and no residual strength tests are planned for this aft boom structure.

The third test (4C) was intended to study the pressures generated by detonation in a water-filled box (Griffis, 1995a). The test articles were stainless steel boxes with the approximate shape and volume of an aft bay. They were filled with water, and an explosive (equivalent to a 30mm HEI) was detonated in the center. Boxes of two different volumes (by 18 percent) were tested; the SPO indicated that the peak pressures are independent of these box volumes. This result was used to justify direct application of the test results from scaled-down test articles to full-scale test articles. The report for this test was not written at the time of the committee's inquiry.

A representative live fire test (Test 4D) was planned for August 1995, when a section of the tank between Frames 5 and 6 was to be filled with water, externally loaded, and hit with a 30mm HEI round (Griffis, 1995a). As of this writing, analyses are being performed to predict the amount of expected damage and the resulting residual strength. No experimental verification of the residual strength is planned.

To evaluate concerns about fuel entering the cockpit and about structural damage that could prevent pilot ejection after a hydraulic ram failure of the forward fuselage tank (F-1), the SPO was planning Test 7 to begin in June 1995 (Griffis, 1995a). This test involves firing either 30mm HEI or API projectiles into a full-size F-1 fuel tank with supporting structure, including the seat back

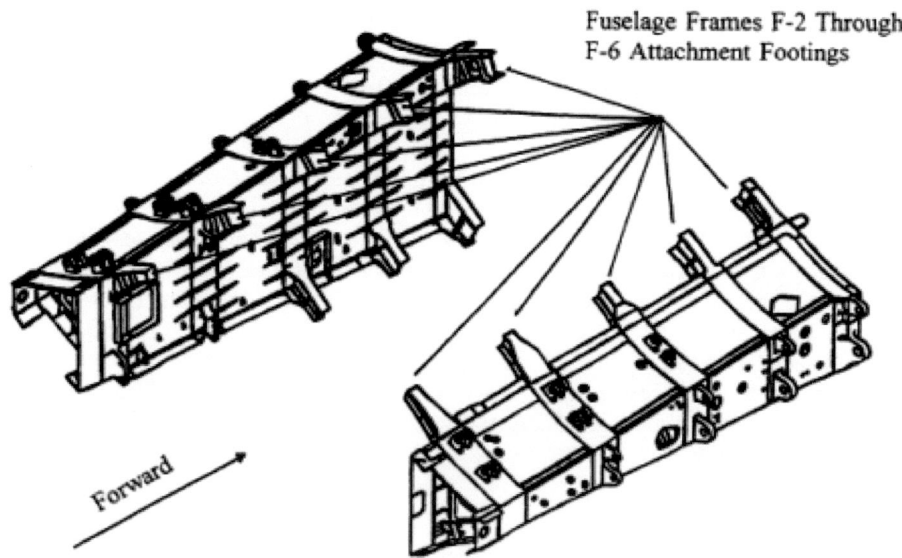

Figure 4-6 Forward boom A-1 fuel tanks. Source: Griffis, 1995b.

bulkhead. Since this test involves single shots to single points in the F-1 tank, the results will only be significant if they are complemented by a thorough analysis of the problem.

It should be noted that no live fire tests are currently planned to determine the damage that would be encountered from a direct 30mm HEI impact on Frame 6, which is common to both the forward and aft sections of the tail boom structure. As was pointed out above, the committee is concerned that failure of this frame could jeopardize flight safety. Likewise, no live fire tests are planned to assess the damage that would occur from a direct hit on the pivot shafts, which provide only single-load-path support of the horizontal tails.

Assessment

The basic F-22 structural design appears to derive substantial battle-damage tolerance from the use of materials with high toughness and the incorporation of multiple load paths. Nevertheless, a major uncertainty is the prediction of damage due to hydraulic ram effects from a ballistic hit. The SPO has recognized this uncertainty and constructed a comprehensive live fire test program to uncover weaknesses in the design. Appropriate hydrodynamic modeling (finite element, finite difference, etc.) would maximize the information extracted from this test program and allow prediction of response for other geometries, fill ratios, and impact kinematics. In fact, the program has already disclosed a weakness in the wing design and corrective measures have been taken (Griffis, 1995a).

Tests 1, 2, 3, 4, and 7 are live fire structural tests, and Test 10 obtains basic penetration data on structural materials. All except Test 4B and material penetration shots in Test 10 involve evaluation of hydraulic ram effects (Griffis, 1995a).

Although a hydraulic ram test is planned for the aft boom fuel tank area, the committee is concerned about the lack of a live fire test shot at Frame 6. Also, there has only been one shot to predict damage from a hit in the dry bay areas of the aft boom. The prediction of damage as well as the remaining residual strength of the boom structure, the aft fuselage frame, and the horizontal tail pivot shafts are all believed to be important. In addition, if it does appear possible that a tail could be lost, it is then important to estimate the risk of aircraft loss.

The committee is concerned that the test specimen planned for Test 4D may not be representative of the aft fuel tank. In particular, the concern is that the absence of fuel on the other side of Frame 5, and the inaccurate representation of Frame 5, could adversely affect the fidelity of the test. The committee believes that hydrodynamic analyses can be used to establish the credibility of the test specimen and to determine if the specimen can represent only the part of the tank between Frames 5 and 6, or if more of the tank needs to be represented.

Suggested Revisions

The committee suggests that the following revisions to the vulnerability assessment program be considered:

- Conduct additional live fire testing to determine the damage that can be expected from a hit in the Frame 6 aft boom attachment area. The Air Force should determine the most critical shot lines (i.e., whether or not they should go through the fuel tank area).
- Expand analyses to predict damage sizes and residual strengths of the aft boom, Frame 6, and horizontal tail pivot shafts after being hit by 30mm HEI rounds. Also, determine the risk of aircraft loss should it be found that loss of a horizontal tail is possible.
- Conduct further analysis of the aft fuel tank (A-1) prior to the conduct of Test 4D. This analysis should be focused on determining the adequacy of the test specimen, with particular emphasis on its ability to simulate accurately the reaction of the entire tank.

Fuel System and Associated Dry Bays

Description and Attendant Vulnerabilities

The F-22 fuel system consists of integral fuel tanks within the aircraft fuselage and wing structure, as shown in Figure 4-7. The fuel system also includes the pumps, valves, plumbing, and components necessary to supply the required fuel flows and pressures to the engines during all flight conditions. Each engine is fed fuel from a common forward and two independent aft feed tank systems. The fuel system is designed to ensure mission completion after one failure and safe aircraft recovery following two failures. All the fuel on board the aircraft is available to either engine via a cross-feed manifold, if no more than a single failure has occurred.

The fuel system on any aircraft, and particularly a fighter, is the single largest nonstructural subsystem on the aircraft and has a large presented area from any threat aspect. If the fuel system is not protected with vulnerability reduction measures, it is also the largest vulnerable area on any aircraft. The primary kill mechanisms of the fuel system are the following (Griffis, 1995a):

- Fire or explosion inside the fuel tanks caused by ignition of the fuel-air mixture in the ullage above the fuel as a result of API or HEI projectiles or other ignition sources.
- Fire in dry bays around fuel tanks caused by projectile or fragment ignition of the fuel spurtback from penetration of the fuel tank.
- Hydraulic ram from projectile or fragment penetration into full or nearly full fuel tanks, which results in fluid shock wave forces that rupture the fuel tank.
- Fuel depletion from leaking or ruptured fuel tanks or fuel lines.

Discussion of elements of the fuel system that are primarily related to structural problems is not repeated in this section.

The ullage spaces are vulnerable to threat-induced fire or explosion unless protection is provided. The F-22 design includes an on-board inert gas generating system (OBIGGS), which replaces air in the ullage spaces with an inert gas (discussed below). In addition, the F-22 has many dry bays that must be protected to prevent fires.

The SPO has addressed the dry bay fire problem by providing fire extinguishing in the main wheel wells and the aft wing attachment bays as well as foam on the top and sides of the F-1 fuel tank. Figure 4-8 depicts these and other vulnerability reduction features. (Fire extinguishing and foam are discussed later in this chapter.)

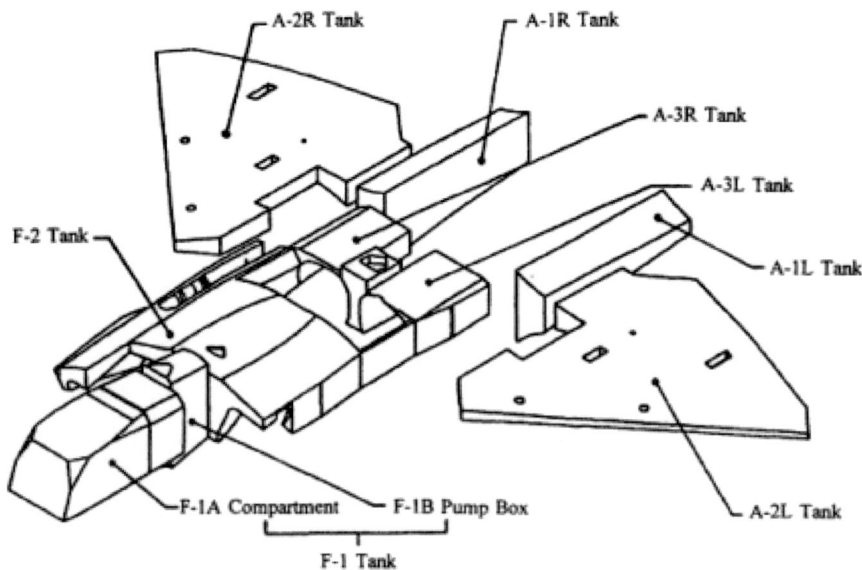

Figure 4-7 Fuel system vulnerability testing. Source: Griffis, 1995b.

A fuel ingestion problem arises when combat damage to a fuel tank allows leakage to be ingested into an engine inlet. For the F-22, this problem has been addressed by having a tailored fuel-burn sequence that will leave the fuselage fuel tanks next to the engine inlets empty when a 60 percent fuel state is reached (Griffis, 1995a). This approach assumes that the aircraft will not see combat until that point in time. Nevertheless, the solution carries with it a degree of risk, and those who fly the aircraft will have to decide if the risk is acceptable. At a minimum, mission planners and pilots should be informed of this risk.

The F-22 fuel transfer lines are located inside the fuel tanks to reduce the chance of their being hit (SPO, 1995a). The transfer lines are also located in inerted fuel tanks. This combination is a good vulnerability reduction design technique. In addition, with no more than a single failure in the cross-feed manifold and feed tank pump, their redundancy allows for full availability of all the fuel within the aircraft to either engine. This cross-feed capability reduces the vulnerability of the system. Flammable fluids that leak are drained overboard through drain holes located in external surface panels along the bottom of the aircraft.

Planned Analyses and Tests

The OBIGGS provides the single largest vulnerable area reduction of any vulnerability reduction feature used on the aircraft. No destructive ballistic tests are planned for this system because previous extensive ballistic tests carried out by the Joint Technical Coordinating Group on Aircraft Survivability (JTCG/AS) have shown that, if the fuel tank ullage is inerted with nitrogen from the OBIGGS to reduce the oxygen content below 9 percent in the ullage fuel-air mixture above the liquid fuel level, no fire or explosion will occur in that area.[4] The committee agrees with the facts in this situation and with the decision not to conduct destructive ballistic testing because it is not necessary.

Dry bay fire protection tests are covered later in this chapter.

Assessment

There is a question of whether OBIGGS generates enough nitrogen to keep the fuel tank ullage spaces inerted at all times. The SPO plans to evaluate this feature by running ground tests in late 1995 on a fuel system simulator to determine inerting performance. This simulator of the complete F-22 fuel system can simulate fuel transfer, slosh, and vibration, and can generally test for proper fuel system functionality before the aircraft actually flies (SPO, 1995b). Sensors will be installed in each fuel tank of the simulator. This will verify that the oxygen measurement sensor determines the concentration of nitrogen-enriched air and activates the OBIGGS. EMD aircraft will have a sensor in each air vent line. Flight test data can then be related to the ground fuel tank simulator data.

Suggested Revisions

The committee has no suggested revisions to the test program.

The lack of testing of a potential fuel ingestion problem and the rationale for it are noted. The committee believes that the operational community (e.g., mission planners and pilots) should be made fully aware that a fuel ingestion risk to the aircraft exists at a fuel state higher than 60 percent.

---

[4] Studies accomplished at the Naval Air Weapons Center in the late 1980s indicated that an oxygen concentration at or below 9 percent provides total fire suppression capability in the fuel tank ullage. These studies agree with tests completed by the Air Force in the 1970s.

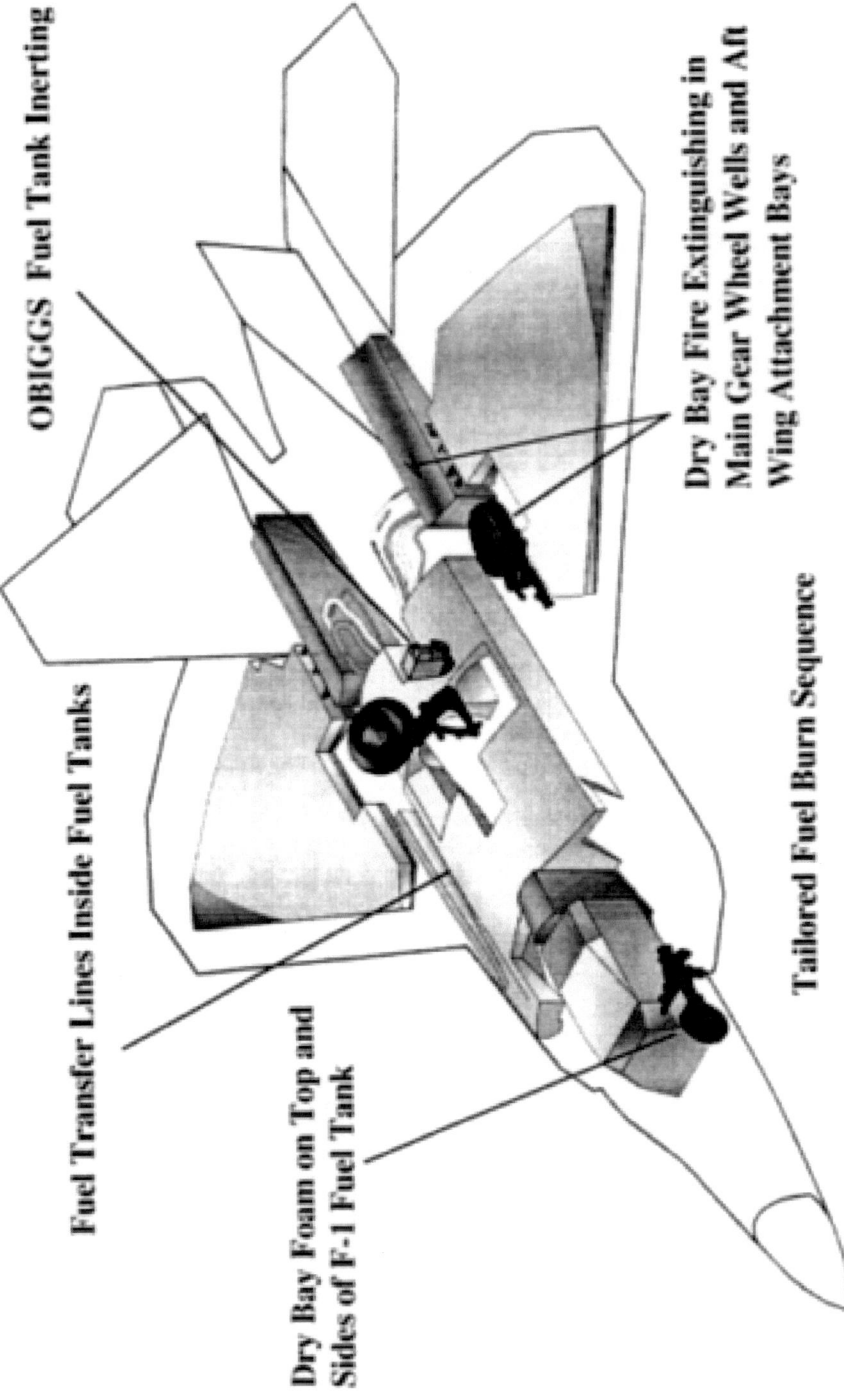

Figure 4-8 Vulnerability reduction features of the F-22. Source: Griffis, 1995b.

Flight Control and Auxiliary Systems

Description and Attendant Vulnerabilities

The term "flight control and auxiliary systems" is broadly construed to encompass those systems necessary to ensure safe, controlled flight. Included are the flight data systems and avionics that generate signals to control the aircraft, hydraulics that actuate the flight control surfaces, and electrical systems that power the avionics. In addition, an environmental control system provides cooling for mission avionics and the cockpit while an on-board oxygen generating system provides enriched breathing air. These latter two systems are necessary for mission completion but not for safe flight.

The air data system and control avionics consist of doubly and triply redundant modules and sensors spatially distributed around the cockpit area. The critical units are air cooled and thus not dependent on the closed-loop liquid cooling system (which uses the flammable fluid, PAO (polyalphaolefin)) required for much of the other avionics on the vehicle. However, coolant is present in many of the relevant avionics bays, and there is the potential for a coolant fire to extend to and disable flight-critical components.

The principal sources of hydraulic and electric power are the two airframe-mounted auxiliary drives (AMAD). Each driven by a different engine, the AMADs consist of a gearbox-mounted hydraulic pump and two electric generators. Additional electric and hydraulic power can be provided by the auxiliary power unit (a small gas turbine engine), which can be started in flight. Each AMAD powers an independent, multiply branched, isolated hydraulic system equipped with leak detection and automatic shutoff features. The electric power system is similarly redundant and fault tolerant. The design intent is that the aircraft be controllable with a single hydraulic and electric power system.

The classic aircraft vulnerability is a single-shot kill of multiple branches of redundant hydraulic, electric, or control systems. Early analysis of the F-22 showed such problems in the electric power distribution centers, the hydraulic system, and the flight control avionics and wiring. Components were relocated and lines rerouted accordingly (Griffis, 1995a).

These systems do, however, continue to present vulnerability concerns. Placement of the AMADs is such that both main hydraulic pumps are collocated near the aircraft centerline, close enough for a single-shot kill of both units. The horizontal tail actuator has been of concern in other tactical aircraft with flying tails because actuator failures can result in hard-over actuation, leaving the vehicle uncontrollable (Griffis, 1995a). This actuator bay does not have fire protection. There is also a fire concern with the aileron actuators and the flight control avionics owing to the proximity of the flammable cooling fluid.

The details of fragment interactions in crowded equipment bays are difficult to predict. So are possible synergistic interactions between systems. An example is the reaction of battle-damaged wiring to sprays of combustible liquids such as fuel, hydraulic fluid, and coolant.

Planned Analyses and Tests

The flight control and auxiliary systems have been subjected to extensive analysis. Live fire testing has been conducted or is planned to explore many of these systems. Test 6 will place shots in the forward fuselage lower and upper avionics bays, AMAD bay, wing leading-edge bay, and main landing gear bay (SPO, 1995a). The aileron bay was shot in Test 11 (Griffis, 1995a).

Test 11 consisted of many shots with 30mm HEI projectiles on a test article representative of the aileron bay. The objectives of the test were to determine (1) the burst radius required to start a fuel-based dry bay fire, and (2) whether the vulnerability analysis overpredicts fires (Holthaus, 1994). The test article had the same structure, materials, and fuel as the EMD configuration. Mock-ups of representative components were included and contained coolant and electrical lines under operating conditions. The test article was subjected to an air flow of 400 knots during the tests.

Eight shots were fired through the dry bay, through the fuel bay, or at the spar in between the bays. In all but one shot, there was no fire (Griffis, 1995a). The shot closest to the spar (at 1.5 inches) ignited a brief fire. Five shots were fired at the electronic warfare box, and 4 of the 5 resulted in fire. The electronic warfare box contained both electrical and PAO lines. Structural predictions were made and compared with results but have not been reported. The results of Test 11 were qualitatively consistent with analysis predictions.

Assessment

Despite subjecting the flight control and auxiliary systems to extensive analysis and testing, uncertainties remain under the current plans. Chief among them is that the current vulnerability analysis does not properly account for the flammable properties of hydraulic and cooling fluids. Both test data and analysis methodology are currently lacking in this area.

Suggested Revisions

Flammability testing of both the PAO coolant and the hydraulic fluid is needed. The SPO's proposed testing of fluid flammability (discussed earlier) will diminish the uncertainties identified above. Thus, this testing should be undertaken. Test data must then be incorporated into models suitable for vulnerability assessment.

Weapons Bay and Ordnance

Description and Attendant Vulnerabilities

The F-22 weapons carriage system includes internal and external weapons, missile launchers, and built-in weapons-loading equipment. In the low susceptibility configuration, all weapons are carried internally. The primary configuration is four AIM-120 missiles, two AIM-9 missiles, and an M61A2 20mm gun.

While weapon vulnerability is always a major factor in overall vulnerability of the aircraft, its importance is amplified by internal storage. This is especially true for the rocket motors that, if ignited when internally mounted, can cause catastrophic damage. The F-22 design also includes the capability to carry weapons externally. Vulnerability analyses of the aircraft to date have not considered external storage because that configuration is not compatible with the combat missions currently defined for the aircraft.

On-board ordnance is vulnerable to fragments impacting the weapon and causing reaction of the high energy components, rocket motor fuel, and high explosives in the warhead. It is also vulnerable to fires in the weapons bay. There have been no special provisions to protect the on-board ordnance from fragments or fire, but some reduction in vulnerability is provided by the surrounding structure of the aircraft, including weapons bay doors.

As noted earlier, vulnerable areas for on-board ordnance are not covered by specifications for the aircraft under the assumption that they are out of the control of the aircraft designer. However, the SPO did cover these areas in mission analysis of the survivability of the aircraft.

The ammunition for the M61A2 20mm gun is electronically primed. Tests on the gun system have shown, for this ammunition, that the ignition of any round in the system will not result in the sympathetic ignition of any other round. Thus a hit on the ammunition, while killing the gun itself, will not result in loss of the aircraft. Therefore, the gun ammunition is not included in the vulnerability assessment.

SUFFICIENCY OF F-22 TESTING PLANS

Planned Analyses and Tests

In the analysis of F-22 vulnerability, it is conceded that the burning or explosive reaction of either the rocket motor fuel or the warhead explosive, when carried internally, would result in loss of the aircraft (SPO, 1995a).

There are no tests planned of the vulnerability of on-board ordnance to fragments or projectiles. The JTCG/AS and the Joint Live Fire Test Program will be relied on to provide estimates of the vulnerabilities of the F-22 due to on-board ordnance for purposes of end-game analyses. It is recognized that these estimates are not complete, and new estimates will have to be included when they are available. This is one case in which the F-22 SPO has not been provided sufficient tools or data to do a completely realistic assessment of the vulnerability of the aircraft (see discussion of tools in Chapter 5).

While there are currently no proposed tests of the ordnance on the F-22, the SPO has suggested that, if additional funding were available, it would consider a series of tests to prove the efficacy of using ablative coatings to protect the internal weapons bays from rocket motor fires (see discussion earlier in this chapter) (Graves, 1995).

Assessment

The committee believes that failure to cover the effects of on-board ordnance in the vulnerability specifications is illogical. The SPO could analyze the implications of on-board ordnance in the vulnerability specifications and have the contractor account for them in the aircraft design (e.g., the structure could provide a degree of protection from fragments). Also, the impact of aircraft design on protection of ordnance and on possible mitigation of the effects of damage should be considered important to survivability of the pilot even if it were not so important to survivability of the aircraft.

While the committee understands the SPO's decision, the vulnerability contribution of the ordnance has not been given adequate attention. For example, insufficient attention has been given to defensive measures such as sensing the inadvertent ignition of an internally carried rocket motor and ejecting the weapon, using ablative material to protect the aircraft from burning rocket motors, or affording greater physical protection to the internally carried rocket motors, warheads, and gun ammunition.

Also, the vulnerability of the ordnance has not been fully established. The JTCG/AS and Joint Live Fire Test Program require further funding to accomplish this task.

Suggested Revisions

The committee fully agrees with the SPO's proposal to establish the efficacy of ablative materials in the weapons bays. Further analysis of the tradeoffs associated with additional ordnance protection or defensive measures such as ejecting burning weapons is suggested.

The JTCG/AS and the Joint Live Fire Test Program should be funded to assure the completeness of data on the vulnerabilities of on-board ordnance.

Engines

Description and Attendant Vulnerabilities

The F-22 is powered by two F119 low bypass ratio, afterburning turbofan engines adjacently mounted at the rear of the aircraft. The F119 incorporates many innovations intended to provide superior performance, maintainability, and survivability (SPO, 1995a). Survivability features include the use of nonburning titanium[5] and a robust, adaptable engine control system. It is the first afterburning engine to be equipped with two-dimensional vectoring nozzles integrated into the flight control system.

The committee believes that dual-engine aircraft, like the F-22, may have an inherent survivability advantage over single-engine vehicles since, while both engines are needed to complete a mission, only one engine is needed for safe flight. However, there are failure modes of one engine (e.g., a disk burst or uncontained engine fire) that will likely result in loss of the adjoining engine and structure and thus in loss of the aircraft.

Many of the F119's technologies have been incorporated into civil or military engines of recent vintage. However, these technologies are not represented in the live fire data base, which mainly contains information derived from tests of older engines. Thus, the accuracy of the empirically based vulnerability analysis of the engine has new uncertainties (e.g., (a) foreign object damage tolerance and fuel ingestion resistance of composite fan stators, (b) failure dynamics associated with ballistic impact and subsequent containment of hollow fan blades, (c) fracture dynamics of new disk alloys and the integrally bladed rotors fabricated from them

---

[5] The F119 engine is to be constructed with titanium "alloy C." The committee understands that the engine contractor has tested this alloy under conditions representative of use in this aircraft, and it will not ignite and burn where more common titanium alloys readily do. Fracture mechanics and fabrication problems have inhibited its use previously. The committee did not independently review these data, but it accepts the contractor's judgment.

in response to ballistic impact, (d) response of the mechanically complex vectoring nozzle to impact damage, and (e) vulnerability of an engine actuation system employing fuel as the working fluid).

Some other vulnerability concerns are raised by the engines' installation. While there is considerable redundancy in the engine control system, both control units and their associated wiring are located on the bottom of the engine, as are all the engine accessories for reasons of maintainability. Thus, all the accessories and controls are vulnerable to multiple fragment impacts on the bottom of the aircraft.

Although the fuel ingestion hazard is considerably reduced once the front tanks are emptied early in flight, fuel remains a concern since there are fuel tanks on either side of the engines. Thus, a projectile shot line from the side that punctures an engine will have first punctured the adjoining fuel tank, admitting fuel to the engine bay at the same time the fuel is heated by hot gas from the engine puncture. The engine bay fire protection system is manually actuated, so prompt action will be required by the pilot in such cases.

Planned Analyses and Tests

Vulnerability analysis of the F119 engines for the F-22 has been carried out with existing models (see Chapter 5) in much the same manner as it has for the other aircraft systems. No live fire testing is currently planned under the F-22 program on the F119 engine or its unique components, although the Joint Live Fire Test Program has an unfunded test program for such engines.

Assessment

Modem engines are relatively vulnerable systems with little history of full-up, full-scale live fire testing. Rather than depend on such testing, engine vulnerability estimates are derived from analysis based on subcomponent results. Much of the data base used in the F119 vulnerability analysis codes stems from the testing of much older engines and components constructed with different materials and design features. Thus, the uncertainty of these calculations must be considered relatively high until they are validated by test data. It should be emphasized that the new technology in the F119 does not necessarily increase the engine's vulnerability Oust the opposite in many cases). It does, however, increase the uncertainty of the vulnerability analysis.

Suggested Revisions

Although some of the needed component testing is included in the 1996 and 1997 plans of the Joint Live Fire Test Program, these tests are not currently funded. The committee believes that these engine-related tests should be pursued, with a focus on F119 components.

Flight Crew

Description and Attendant Vulnerabilities

Design of the F-22 includes significant capability for the survivability of the pilot. Three primary systems must be addressed: (1) the pilot, (2) the ejection seat and escape system, and (3) the life support system. In vulnerability analyses, the first two are considered in the context of a kill and the last in the context of mission abort.

The pilot is vulnerable to fragments, blast, fire, laser damage to the eyes,[6] toxic fumes, chemical and biological weapons, spall, ricochet, secondary debris, secondary damage resulting from explosion of fuel in the F-1 fuel tank, failure of the ejection system, and so forth.

Fire in the cockpit area would certainly, if uncontained, result in loss of the aircraft. However, fire should not prevent crew ejection as long as critical components of the ejection system are not adversely affected (e.g., the seat's ballistic components or components that affect ejection sequence timing). Cockpit fires could result from ignition of PAO fluid in surrounding electronics, from the explosive elements of the ejection system, or from fuel leaking from ruptured tanks in the forward area. Vulnerabilities to fire are discussed elsewhere in this chapter.

Associated with fire damage to the aircraft are toxic fumes and smoke, which could be drawn into the cockpit from nearby fires or other damage. The life support system helps to protect the pilot from this threat.

---

[6] With respect to laser damage to the pilot's eyes, the pilot's visor is considered to provide adequate protection (Giorlando, 1995).

Planned Analyses and Tests

Flight crew vulnerability to fragments, including secondary spall and ricochet, is based on standard JTCG/AS tables and determined as a vulnerable area, contributing to the overall vulnerable area of the F-22 (SPO, 1995a). While there is no special protection for the pilot in the design, some shielding is provided by surrounding structure and electronic modules. Modeling is used to assess the vulnerable area in the presence of the surrounding structure and modules.

The committee received indications from the SPO (Ogg, 1995) that pilot protection methods had been evaluated. The SPO determined that the protective methods considered (e.g., Kevlar) would be only marginally successful. In addition, the added weight of the material (over 100 pounds), while decreasing flight crew vulnerability, had a greater negative effect on overall system survivability than the relatively small positive contribution of the protective material.

While pilot vulnerability to blast is not explicitly dealt with, it is implicitly and adequately handled through the blast vulnerability assessment of the aircraft. Vulnerability of the forward fuel tank to fire or explosion, and the potential impact on pilot or escape system, is discussed elsewhere in this chapter.

The committee is not aware of any planned live fire testing of crew or escape system vulnerability. Rather, vulnerability of these components is calculated by modeling, and a single fragment hit on these areas is conceded as loss of the aircraft.

Assessment

Owing to time constraints, the committee was not able to assess the analytical methodologies being applied to flight crew survivability within the vulnerability community. The committee accepts the SPO's belief that it is very difficult to protect flight crews from modem threat systems without severely impacting overall survivability. Nevertheless, the committee believes that efforts to reduce flight crew vulnerability should continue, both by the F-22 SPO and the JTCG/AS, because of the flight crew's contribution to overall F-22 vulnerable area.

Suggested Revisions

The committee has no suggested revisions to the test program. However, the F-22 SPO and the JTCG/AS should emphasize their continuing efforts to develop improved methodologies for reducing flight crew vulnerability.

Fire Protection Systems

Description and Attendant Vulnerabilities

The fire protection systems detect, isolate, contain, and extinguish fires and suppress explosions. Advanced optical fire detection sensors are included for rapid fire detection. In addition to active fire detection and extinguishing components, fire safety is enhanced by ventilation and drainage of flammable fluids and fuel tank inerting.

Combustibles.

Fuel tanks are separated from adjacent compartments by liquid-proof and vapor-proof barriers. Fuel tanks located adjacent to the engine and auxiliary power unit (APU) compartments and the cockpit are separated from these compartments by a second liquid-proof and vapor-proof barrier in addition to the barrier provided by the fuel tank compartment. Environmental control system bleed air ducts are insulated as required to limit the external duct surface temperatures to a maximum of 700°F within the engine compartments and 500°F in dry bays. Steel and titanium plumbing or the equivalent are used in fire zones. When flight-critical flammable fluid lines run through fire zones, the lines are shrouded and the shrouds are vented and drained. Compartments containing flammable fluids or reservoirs, and those adjacent to fuel tanks, are ventilated at 1 to 3 changes per minute to prevent accumulation of flammable vapors (SPO, 1995a).

Fire Detection.

An optical fire detection system provides fire detection capability in the engine and APU compartments. It monitors ultraviolet radiation produced by burning hydrocarbon fuel to sense the presence of a fire. The optical fire detection capability is provided by eight optical sensors located in each engine compartment and four optical sensors in the APU compartment. The fire protection module located in the integrated vehicle subsystem controller monitors and processes the status signals produced by the optical sensors to provide fire and fault information (SPO, 1995a).

A thermal overheat detection system operates along the engine high-pressure bleed ducts and reports overheat conditions. Dual-loop, discrete thermal detector units are installed adjacent to the high-pressure bleed ducts in each engine compartment to detect leaks that may impinge on critical structure or components or constitute an ignition source. Coverage is divided into three zones. Both left and right bleed zones provide detection within the respective engine compartments for the sections of duct between the engine and the bleed air manifold and along the engine anti-icing duct. The center bleed zone provides detection from the bleed air manifold to the primary heat exchanger (SPO, 1995a).

Dry Bay Fire Protection.

Fire detection and suppression is provided for the left and right main landing gear wells and the left and right wing attachment bays aft of the main landing gear wheel wells. Detection and suppression is provided by a dry bay fire protection unit that combines an infrared optical fire sensor and a pressurized Halon-filled cylinder in an integrated unit (an agent to replace Halon is being sought). The extinguishing agent is automatically discharged when a fire is detected. For fire containment and control, firewalls separate the engine compartments, engine nozzles, and APU compartment from adjacent compartments. The firewalls are designed to prevent flame penetration for 15 minutes when subjected to a 2000°F flame. Critical components are fire hardened to prevent damage from exposure to flame or fire-generated thermal energy. A fire extinguisher provides suppression capability to either engine or APU compartments. The extinguisher contains approximately five pounds of Halon (a one-shot system) (SPO, 1995b).

Dry bay fire remains the largest vulnerability contributor of the fuel system. As a general rule, dry bay fire occurs anywhere a fuel tank, component, or line is adjacent to an internal dry bay that is not protected by fire extinguishing. Fire protection is included in 15 separate bays. Foam is located in strategic areas around the F-1 fuel tank (behind the pilot) where the majority of hit-initiated fires could occur (SPO, 1995a). These areas were unique in the design because they were also unoccupied by components and could easily be filled with the foam.

Fuel System Explosion Suppression.

The fuel system is designed with functional redundancy and component separation so that a single hit will not interrupt fuel to both engines. The current fuel usage schedule burns fuel from selected tanks, which are assumed to be emptied early. The fuel ingestion kill mechanism associated with this assumption was discussed earlier in this chapter. OBIGGS (discussed earlier) protects against ullage explosions.

Planned Analyses and Tests

Dry Bay Fire Extinguishing.

The fire detection sensors are to be tested under laboratory conditions. The fire detection sensor's field-of-view analyses are completed. The plan for Test 6 details the ballistic dry bay fire tests that will evaluate the dry bay fire protection system (SPO, 1995b). The areas of interest for Test 6 are illustrated in Figure 4-9. It is expected that the Halon replacement agent (see below) will be less efficient. Therefore, containment bottles may need to be resized accordingly with some size and weight increases.

Evaluation Plan for Halon Replacement.

The Halon Replacement Program for Aviation covers more than just the F-22 program. An alternate fire suppression agent for Halon has been identified for use in aircraft dry bays and engine cells. The replacement program involves a three-phase effort for both applications (SPO, 1995b):

- Phase I studied the operational parameters that affect the amount of agent needed for each fire environment. The four most significant parameters were found to be surface temperature, air temperature, fire location, and fuel type.
- Phase II was an operational comparison of three selected agents. Testing was conducted to screen and compare performance data on alternative agents, and one agent (known as HFC-125) was selected.
- Phase III is under way now to establish design criteria methodologies. A product of Phase III will be design equations for use in sizing suppression systems that use the new agent.

The F-22 program will not need to evaluate the Halon replacement agent. It will test and evaluate the installed F-22 dry bay extinguishing system.

Dry Bay Foam.

The SPO feels confident that the characteristics of dry bay foam are well understood—foam is expected to provide protection only against missile fragments and small caliber API threats. For this reason, no tests are planned to evaluate the material specifically (SPO, 1995a).

Synergistic Effects.

Synergistic events, such as fires caused by arcing electrical power wires being sprayed by environmental control system coolant or hydraulic fluids, are poorly understood. These events must be recognized as a potential kill mechanism with a high degree of uncertainty. Test 6 will examine the synergistic effects of PAO liquid coolant fluid, reduced flammability hydraulic fluid, aircraft fuel, and electrical power in protected and unprotected, cluttered aircraft dry bays. Test 6 will also examine fire detection and extinguishing capability in protected aircraft dry bays of the F-22.

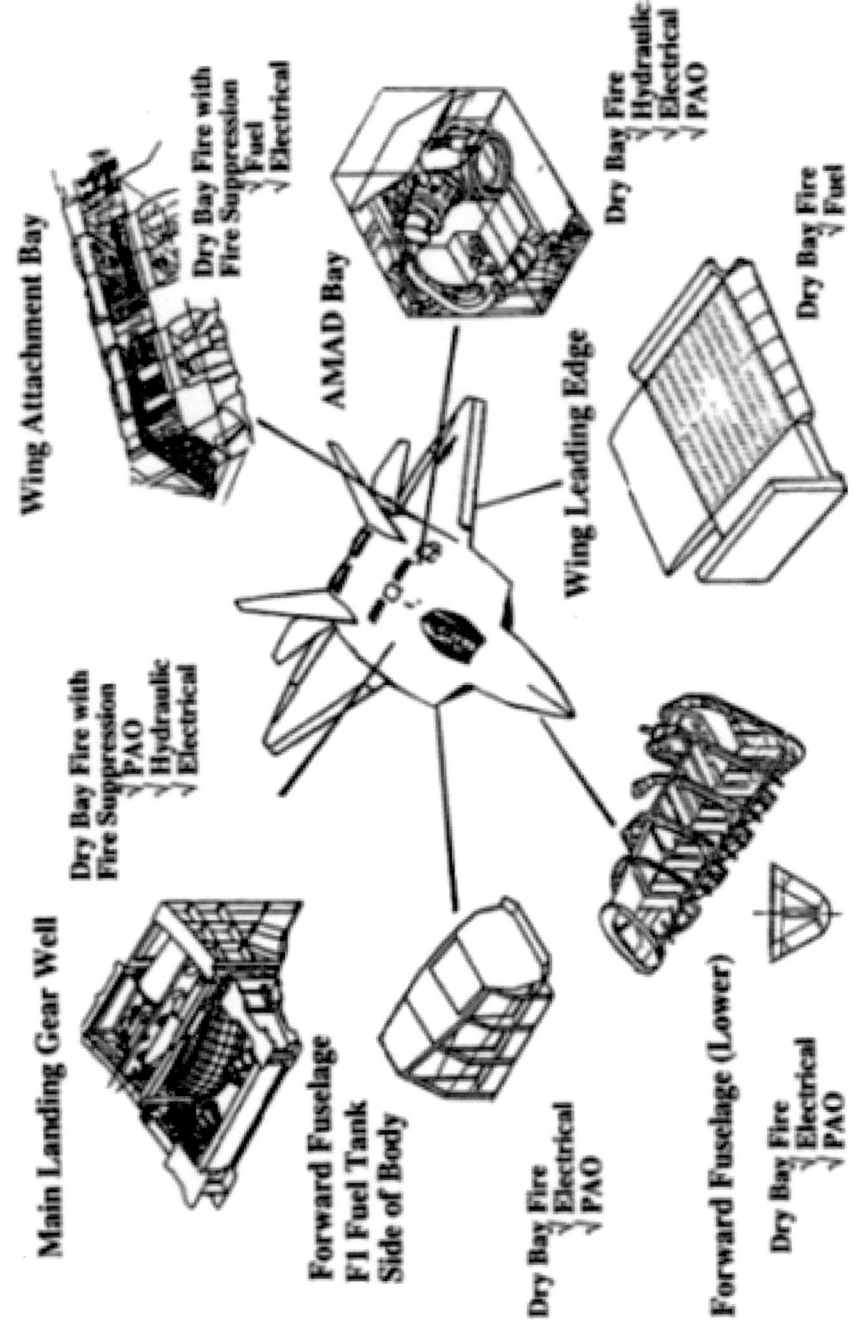

Figure 4-9 Areas of interest for Test 6. Source: Giffis, 1995b.

Assessment

The infrared sensors used in the fire detection system are a strength of the system. They are dual-channel sensors, each with a 90-degree field of view. Their response time is between 5 milliseconds to detect a fast growing fire and 5 seconds to detect a slow growing fire. Their wiring logic incorporates power interrupt detection. Accessibility for maintenance was a major consideration in design installation. The integrated vehicle subsystem controller is designed to provide essential redundancy for cockpit indication of any existing fire (SPO, 1995b).

Synergistic Effects

Synergistic effects are those in which the initial damage to one subsystem may result in effects that cause further damage to one or more other subsystems. The damage may continue until the aircraft is killed. Some call this phenomenon "cascading effects."

For example, a shot may damage electrical wiring and pass on into a flammable fluid container (e.g., hydraulic fluid reservoir or lines) without initiating a fire from the incendiary effects. However, the fluid may then leak into sparking wires from the severed electrical cables, which could result in a fire. The F-22 has areas where these synergistic effects could occur.

Test 6 is a comprehensive and systematic way to evaluate and optimize the effectiveness of dry bay protection. Test article frames are designed to withstand multiple ballistic impacts, can be easily reconfigured after each test, and contain representative bay sizes. Environmental control system airflow will be used when necessary. Electrical wiring with operating voltage and currents will be incorporated, significantly adding to the validity of the tests. Test 6 is a valuable subsystem type test that cost-effectively uses other-than-flight hardware and allows optimization should problems be uncovered.

Test 11 was discussed earlier. In Test 6 and Test 11 together, a large number of shots (approximately 70) have been made or are planned. Even though the test articles do not contain all of the fluid lines and electrical wires, but only the major ones, it seems likely that this number of tests will allow the development of a robust methodology that can be extrapolated to estimate differences under entirely realistic conditions.

Test 6A will examine the synergistic effects of pressurized PAO coolant lines and PAO-cooled avionics modules and the adjacent powered electrical wiring in the F-22 forward fuselage lower avionics bays. The SPO indicated that it may use the prototype air vehicle fuselage in Test 6A and will include all sources of PAO fluid and electrical wiring in this test (Griffis, 1995a). Greater realism in this type of test has merit. The test program is quite thorough, but the committee believes that the prototype air vehicle fuselage should be used in Test 6A in order to achieve greater realism.

Suggested Revisions

The committee urges use of the prototype air vehicle fuselage in Test 6A but has no other suggestions for the planned test program.

ADDITIONAL OBSERVATIONS

The F-22 SPO has devised and implemented a program consisting of vulnerability reduction, design assessment, and testing that is founded on (a) comprehensive analysis of the vulnerabilities of the aircraft, (b) identification of areas of uncertainty (as discussed in the introduction to this chapter), and (c) tests and assessments required to address these uncertainties. The committee believes that this program is well conceived and sufficiently realistic to support the request for the waiver of the live fire test law. If modified as suggested earlier in this chapter, the program will be strengthened as the F-22 proceeds with EMD and initial production.

However, vulnerability assessment is a complicated and difficult problem. Therefore, the committee carefully considered the need for continued testing of the vulnerability of the F-22, beyond the current scope. In the process, the committee held detailed discussions with the vulnerability and lethality assessment communities.

At China Lake (Navy and Air Force personnel and a consultant to OSD were present), members of the committee were briefed on recent and planned live fire test programs for Navy aircraft (Tyson and Wise, 1995). The committee has considered the Navy's assessment methodology in comparison to that of the Air Force and the F-22. While there are some philosophical differences between the Air Force and the Navy, there is a great deal in common where aircraft are concerned.

The opinions below are held by the China Lake team and were not disputed in the meetings by the Air Force representative present.

- Given a complete aircraft, more information can be extracted by testing its major parts than by testing the entire vehicle at one time.
- It is possible that additional data could be obtained from such testing of the F-22 that could result in revisions to the aircraft design to further reduce its vulnerability. Similar large assembly testing has been conducted and is planned for variants of the Navy's F-18 even though the China Lake team does not expect to discover unanticipated outcomes.

The committee grappled with the question of whether to recommend similar testing for the F-22. The paradox seems to be the question of recommending tests that are not expected to yield unanticipated results, especially for such an expensive aircraft. The Navy's rationale seemed to be that ongoing testing makes sense in that something is always learned, vulnerability assessment methodologies are improved, models and other tools are evaluated, and, like the industrial base, the live fire test base cannot be allowed to wither.

After much deliberation, the committee agreed that the Navy's approach made sense for the Air Force's F-22. This fighter will be in the inventory for decades and can be expected to undergo an evolution that includes other missions and new configurations. That life cycle dictates a continuation of live fire testing.

Accordingly, the committee believes that the Air Force should plan for expeditious vulnerability assessment testing of the F-22 similar to that being conducted or planned for variants of the F-18. In particular,

- As soon as a source of large assemblies can be identified (e.g., from a damaged aircraft or other test hardware that is representative of production aircraft, including assemblies from early production or one of the nine dedicated test aircraft), these assemblies should be provided to the vulnerability assessment community. The committee recognizes that such assets may not become available until after production begins.
- It is assumed that the most useful information could be derived from the resulting test specimen if it is tested for vulnerability in major subassemblies rather than as a complete system configured for combat.
- The testing should be directed at (a) verifying predictions derived from the current live fire test program and the models used, and (b) testing the effects on the overall vulnerability assessment brought about by configuration and mission changes through the years.

It is recognized that the results of such testing will likely not impact the design of the initial production aircraft of the F-22. The committee believes that this level of risk is acceptable in view of the comprehensive nature of the ongoing program modified to accommodate the revisions suggested by the committee. However, the results of such tests would certainly influence future production blocks or modifications of the F-22 to perform future missions. Techniques for repairing battle damage to the F-22's new composite materials and new systems could be verified. Vulnerability assessment tools might also be improved.

CONCLUSIONS

Adequacy of F-22 Threat Definition and Replication

The committee accepts the threat environment defined for the current mission of the F-22. The assumed API and HEI rounds are reasonable replications of air-to-air, cannon-fired threats. The two discrete fragment sizes are representative of the fragments from the spectrum of warheads likely to be encountered. However, for some classes of warheads (e.g., annular or focused blast fragmentation), the kill mechanism that involves dense multiple fragment impacts may be important for the F-22 and should be considered in future analyses and tests.

Overall Sufficiency

The Air Force and its contractors responsible for the design of the F-22 have incorporated a large number of features in the design of the aircraft that will reduce its vulnerability. These features include a structural design with largely multiple (redundant) load paths, inerted fuel tanks using OBIGGS, dry bay foam, double-walled barriers between the cockpit and fuel, redundant fuel pumps and cross-feed between tanks, multiply redundant electric power, redundant hydraulics and flight control actuators, separated triple redundant air-cooled mission computers, fault tolerant avionics, dual engines, and engine blade containment.

In addition, the Air Force and its contractors have performed a detailed vulnerability analysis to determine the vulnerable area of the F-22 using revised versions of standardized computer models. They assessed the uncertainties in the analysis and then constructed a comprehensive live fire test program to address these uncertainties and validate several of the design features.

The committee reviewed this overall assessment program and suggested specific actions by the Air Force and others to alleviate some concerns that the committee has. Given the F-22's current counter-air mission, the program is sufficiently realistic to support the requested waiver. With the committee's suggested actions and the implementation of possible corrective measures as a result of the program findings, the committee concludes that the Air Force program will be strengthened as the F-22 proceeds with EMD and initial production.

Specific Actions

The specific actions suggested by the committee are listed below. For easy cross-referencing, the actions are ordered as in the evaluation section of this chapter.

Structure and Integral Fuel Tanks

- Conduct additional live fire testing to determine the damage that can be expected from a hit in the Frame 6 aft boom attachment area. The Air Force should determine the most critical shot lines.
- Expand analyses to predict damage sizes and residual strengths of the aft boom, Frame 6, and horizontal tail pivot shafts after being hit by 30mm HEI rounds. Also, determine the risk of aircraft loss should it be found that loss of a horizontal tail is possible.
- Conduct further analysis of the aft fuel tank (A-1) prior to the conduct of Test 4D. This analysis should be focused on determining the adequacy of the test specimen, with particular emphasis on its ability to simulate accurately the reaction of the entire tank.

Fuel System and Associated Dry Bays

- Make the operational community fully aware that a fuel ingestion risk to the aircraft exists at a fuel state higher than 60 percent.

Flight Control and Auxiliary Systems

- Conduct the tests and analyses proposed by the F-22 SPO on the flammability of coolant and other fluids and the attendant vulnerability of the aircraft.

Weapons Bay and Ordnance

- Undertake the analysis and test, proposed by the SPO, of ablative materials in the weapons bay. Also, conduct further analysis of the tradeoffs associated with additional ordnance protection or defensive measures.

- Fund the JTCG/AS and the Joint Live Fire Test Program to assure the completeness of data on the vulnerabilities of on-board ordnance.

Engines
- Fund the proposed Joint Live Fire testing of F 119 engine components to alleviate the paucity of testing against those components.

Flight Crew
- Emphasize continuing efforts by the F-22 SPO and JTCG/AS to develop improved methodologies for reducing flight crew vulnerability.

Fire Protection Systems
- Use the prototype air vehicle fuselage in Test 6A in lieu of a mock-up. (The Air Force has considered using this fuselage for Test 6A.)

Additional Action

As discussed above, the committee believes that continued testing of the vulnerability of the F-22 is desirable. This testing should be conducted expeditiously against representative production hardware in the form of large assemblies when they become available. These tests should be similar to those being conducted or planned for variants of the Navy's F-18.

The committee recognizes that large assemblies from a damaged aircraft or other production-representative hardware, from any source, may not become available until after production begins, and the subsequent recommended testing will likely not impact the initial F-22 design. The committee considers this arrangement to be acceptable. However, test planning should begin soon. The use of this hardware and the results of these tests would influence future production blocks or modifications of the F-22 to perform future missions. Additional positive consequences may well include (a) verification of techniques for repairing battle damage to the F-22's new composite materials and systems, and (b) improved vulnerability assessment tools.

REFERENCES

Ball, R.E. 1985. The Fundamentals of Aircraft Combat Survivability Analysis and Design. New York: American Institute of Aeronautics and Astronautics, Inc.

Giorlando, J. 1995. High Power Microwave. Presentation to the Committee on the Study of Live Fire Survivability Testing for the F-22 Aircraft, F-22 System Program Office, Wright-Patterson Air Force Base, Ohio, January 19.

Graves, J.T. 1995. National Research Council Questions on Live Fire Test. Memorandum from Deputy Director, F-22 System Program Office, to Mike Clarke, Commission on Engineering and Technical Systems, National Research Council, February 14.

Griffis, H. 1995a. Live Fire Test Program. Presentation by to the Committee on the Study of Live Fire Survivability Testing for the F-22 Aircraft, F-22 System Program Office, Wright-Patterson Air Force Base, Ohio, January 19.

Griffis, H. 1995b. National Research Council Questions on Live Fire Test. Memorandum to National Research Council, April 28.

Hinton, W.S. 1994. Threat, Mission, and Operational Requirements for the F-22. Presentation to the Committee on the Study of Live Fire Survivability Testing for the F-22 Aircraft, National Academy of Sciences, Washington, D.C., December 21.

Holthaus, T.M. 1994. Live Fire Test No. 11, F-22 Live Fire Test Plan, Live Fire Test of Aileron Dry Bay Fire Test Program. Wright-Patterson Air Force Base, Ohio: Wright Laboratory.

Ogg, J. 1995. Vulnerability Program Overview. Presentation to the Committee on the Study of Live Fire Survivability Testing for the F-22 Aircraft, F-22 System Program Office, Wright-Patterson Air Force Base, Ohio, January 19.

SPO (F-22 System Program Office). 1993. Live Fire Test 4A Report. Memorandum to Roland Yancey, Boeing Aircraft Company. Memo L8932-AL93-059. Wright-Patterson Air Force Base, Ohio: F-22 System Program Office. December 15.

SPO. 1995a. Combat Survivability—F-22 Live Fire Test Program. Presentation to the Committee on the Study of Live Fire Survivability Testing of the F-22 Aircraft, F-22 System Program Office, Wright-Patterson Air Force Base, Ohio, January 19.

SPO. 1995b. National Research Council Questions on Live Fire Test. Memorandum to National Research Council, March 23.

Tyson, H., and T. Wise. 1995. Live Fire Test Program for the F/A-18E/F and V-22. Presentation to the Committee on the Study of Live Fire Survivability Testing for the F-22 Aircraft, Naval Air Warfare Center, Weapons Survivability Laboratory, China Lake, California, February 21.

5

Vulnerability Assessment Tools

Live fire testing can produce data that, when used in conjunction with models having predictive capabilities, will be useful in extending vulnerability assessment to a much greater range of conditions than can practically be tested. Providing data (i.e., input) for predictive analyses is essential because such analyses must be relied on to account for the probabilistic nature of the threat and the many operating conditions during which a combat aircraft may be hit.

Considering the importance of vulnerability assessment tools, in particular the extent to which they complement live fire testing, a review of the array of tools (e.g., documents, data bases, and models) is mandatory for a complete evaluation of F-22 live fire testing. Such a review is the objective of this chapter.

ROLE OF TESTING, MODELING, AND DATA BASES IN VULNERABILITY ASSESSMENT

The objectives of vulnerability assessment are to identify both mission and vehicle kill mechanisms and to estimate quantitatively the robustness of the aircraft to hits from relevant threats. This quantitative information is then used by the designer to produce an optimized vehicle design that takes into account the full range of performance requirements as well as metrics like vulnerability and susceptibility. Cheap kills would be eliminated whenever possible. Accurate vulnerability assessment requires a balance of testing and modeling, aided by information in established data bases.

Modeling is integral to quantitative vulnerability assessment since it is only in this manner that large numbers of threat-target interactions can be examined. The term "modeling" is used widely here to encompass analysis and numerical simulation based on mathematical approximation of structural and fluid mechanics, combustion, detonation, and other pertinent phenomena, as well as statistical bookkeeping.[1]

---

[1] Accurate vulnerability assessment of complex aircraft requires models at many levels: models of basic physical processes (e.g., fuel spray ignition, composite damage, and hydraulic ram), models of subsystem behavior (e.g., fault trees, response of hydraulics to a severed line, and wing

Currently, the models used by the vulnerability assessment community depend on approximations rooted in empirical observations. Thus, models may be improved both by more representative mathematical approximation and by more accurate and complete data. Models must be extensively tested to establish accuracy and limits to applicability.

> The report *Vulnerability Assessment of Aircraft* (NRC, 1993) gives a detailed discussion of the steps in the process of vulnerability assessment. Since that discussion represents the general approach taken by the F-22 SPO, excerpts from it are reproduced in Appendix D to provide interested readers with additional information.

Testing is essential to aircraft vulnerability assessment. Testing is required for several reasons: (a) to establish the essential relations used in modeling (e.g., fuel flammability and composite failure criteria); (b) to validate models; (c) to verify subsystem response to damage (e.g., hydraulic system response to battle damage, ordnance response to projectile penetration, fuel tank response to internal detonation); and (d) to assess the response of major subassemblies or a complete vehicle to threat damage (e.g., to audit the modeling, investigate hard-to-model interactions, and identify failures not predicted by the modeling). Only the last level is controversial, mainly owing to the difficulty of adequate ground simulation of complex flight conditions and the expense it entails.

In the committee's opinion, perhaps the most important use of large-scale testing is verification. Current vulnerability modeling is, at best, an art of estimation. Therefore, testing of complete subassemblies is needed to assess the fidelity of the modeling (i.e., to verify that the models produce acceptable results). This testing must be done in a very judicious manner because, unlike an armored vehicle, an aircraft is a relatively fragile structure, easily destroyed in testing.

Vulnerability test results cannot stand alone. They must be interpreted through modeling to assess the quantitative impact of the test results on overall aircraft vulnerability. Models synthesize the results of discrete tests to make predictions. If the models that are used to replicate the test results are faulty, then the predictions made by the models may be incorrect. Verification, validation, and authentication of models are important steps in the vulnerability assessment process. Empirical models are only as good as the test data on which they are based.

Data bases play a distinct role in vulnerability assessment in that they form the institutional memory that bridges specific systems, prevents repetitious testing, and avoids the mistakes of the past. As with modeling and testing, vulnerability data bases exist on several levels: data bases of constitutive properties (for use in

---

response to a damaged spar), and bookkeeping models to account for aircraft system response to component damage. Definitions of various types of models appear later in this chapter.

VULNERABILITY ASSESSMENT TOOLS 85

models), data bases of engineering practice (for military specifications), and data bases of battle damage (for lessons learned). This information provides both specific data and general guidance for the vulnerability engineer.

Accurate vulnerability assessment requires a careful balance of testing and modeling. Neither is sufficient in itself for a modem weapon system. The following sections address the assessment tools currently available to the F-22 community.

DOCUMENTATION

Many documents have been produced by organizations within the DoD, including the Joint Technical Coordinating Group on Aircraft Survivability (JTCG/AS) and others, that discuss the design of aircraft to reduce their vulnerability to various types of threats. Prime examples include those addressing the use of inerting gas systems in empty fuel tanks and ullage areas, the design of damage-tolerant structures by incorporating dual load paths, and the separation of critical components. Some of these documents were reviewed by the committee; they are listed below to provide the reader with an impression of the type of information that is available.

- Military Standard, Survivability, Aeronautical Systems (For Combat Mission Effectiveness) (DoD, 1986);
- Military Standard, Aircraft Nonnuclear Survivability Terms (DoD, 1981a);
- Military Standard, Requirements for Aircraft Nonnuclear Survivability Program (DoD, 1981b);
- Military Handbook, Survivability, Aircraft, Nonnuclear, General Criteria—Volume 1 (DoD, 1982);
- Military Handbook, Survivability, Aircraft, Nonnuclear, Airframe—Volume 2 (DoD, 1983a);
- Military Handbook, Survivability, Aircraft, Nonnuclear, Engine—Volume 3 (DoD, 1983b); and
- Aircraft Fuel System Fire and Explosion Suppression Design Guide (Mowrer et al., 1990).

The committee noted that, although there are numerous military standards, handbooks, and design guides available, many of these documents are 10 or more years old and have not been updated since they were developed and promulgated. The committee believes that many of these documents need to be updated and improved.

DATA BASES

Designers are aided by vulnerability engineers who use vulnerability assessment models to identify and address system vulnerabilities. Vulnerability engineers and live fire test planners use vulnerability assessment models to help identify areas of the aircraft that require live fire testing. Both processes depend on confidence in the data bases that support the models.

The more that tests are conducted on a certain component, the better the understanding of what that component's damage and failure modes will be. If a statistically significant number of tests are run, the designers, engineers, and testers will have greater confidence in the resulting data base. This example extends to all the components on the aircraft and their data bases.

The Joint Live Fire Test Program managed by JTCG/AS has done some of this testing in the past, and the results have been very beneficial. Unfortunately, the program has been significantly underfunded. This situation has resulted in insufficient component tests being accomplished to allow confidence in the data bases for existing components and materials. In addition, there are many new composite materials, engines, stealth techniques, weapons, and other advances that will require significant additional testing to ensure that the damage and failure modes are understood and confidence in the data bases is warranted.

A building block vulnerability testing approach (i.e., building up from materials to components to subsystems to major subassemblies), with enough tests to produce statistically meaningful results, is necessary to develop the knowledge base to a point where designers, engineers, and testers will have high confidence in the data bases. This effort would generate results that could be useful for all aircraft programs.

The committee believes that this approach could provide significant savings in the live fire test program for individual aircraft. Live fire testing would still have to be accomplished for each new aircraft because each design is different. However, less testing would be necessary for each new aircraft because the confidence level in the supporting data bases would be higher.

MODELS

Modeling is essential for the analysis and assessment of vulnerability. The virtually infinite possibilities for threat-target interactions (e.g., types of threat, intercept geometries, and target conditions) demand a modeling capability to extend the limited number of experiments and tests that can be conducted. Likewise, the design for reduced vulnerability requires an understanding of the processes involved that is reflected in the ability to model.

VULNERABILITY ASSESSMENT TOOLS

> **DEFINITIONS**
>
> The terminology for the extended set of models relating to vulnerability analysis is not uniform. The following set of definitions is deemed reasonable by the committee.
>
> ***Model***: A mathematical construction that describes a physical process or a complex sequence of processes.
>
> ***Closed-Form Model***: A model that involves a mathematical equation to describe the phenomenon (e.g., $F = ma$).
>
> ***Deterministic Model***: A model that produces a definite result for a given situation rather than a probabilistic estimate (e.g., Ohm's law).
>
> ***Empirical Model***: A model that relates a rather complex physical event to a simple equation by a curve-fitting process, with little attempt to describe the actual physical mechanisms involved. (Penetration equations are often of this type.)
>
> ***Encounter Model***: A model that produces an estimate of the probability of kill for encounters between the target and a munition.
>
> ***Numerical-Analysis Model***: A model that requires extensive calculations to derive its results (e.g., a finite-element model).
>
> ***Phenomenological Model***: A model of a physical process that is subordinate to a complete analysis of vulnerability (e.g., a model of the penetration capabilities of a bullet).
>
> ***Probabilistic Model***: A model that describes the process by parameters such as probability of occurrence, mean value, and variance (e.g., an unbiased coin gives "heads" half of the time).
>
> ***Stochastic Model***: A probabilistic model that uses repetitive calculations with random sampling (also called a Monte Carlo model).

It is important to remember that the modeling process involves continual iteration between experimentation and calculation. As in other scientific endeavors, the model represents the hypothesis to be tested by the experiment, and the experiment represents real data upon which to build or modify the model.

Models used for vulnerability assessment are developed both by joint organizations such as the JTCG/AS and by the services. For example, the JTCG/AS is developing a physics-based model for use in predicting fires in dry bays adjacent to fuel tanks. There are plans to expand the model (e.g., to encompass flammable fluids other than fuel and to account for ullage explosions) as funds are made available (Lauzze, 1995). The Air Force has also developed some models for its specific vulnerability assessment of the F-22 (see discussion below).

Phenomenological Models

It is no exaggeration to say that the validity of an encounter model is only as good as the phenomenological models that it employs. These models can be in closed form or can be elaborate numerical-analysis models.

Since the processes involved tend to be quite complicated physically, the closed-form models are often empirical fits to experimental data, and many times they must be expressed in probabilistic terms. Extensive experimentation is implied, both to cover the range of parameters involved and to develop a representation that has statistical validity. Experimentation has been the main approach for obtaining penetration and component-damage data. In the committee's judgment, the vulnerability community has unfortunately not taken full advantage of advances in finite-element and hydrocode analysis—even though these tools would allow a more complete analysis and a considerable reduction in testing. There seem to be two reasons:

- A lack of basic data on the properties of materials that are subjected to extremely high pressures and high rates of loading.
- The large computational times required to calculate individual cases for the deformation of complex structures.

The committee believes that the vulnerability community could do much more to use these advanced tools by (a) developing the necessary data on the materials (especially composite materials), (b) making the lengthy calculations for important instances of damage to structures, and (c) using the results to calibrate and validate simpler empirical models. There is also a need to develop codes using these tools so that the combination of stress, hydrodynamic, and thermal effects can be considered.

The notion of using finite-element analysis to solve complex and combined stress, fluid, and thermal problems is not a remote future option. This type of modeling is already being used by the designers of nuclear weapons and the aerospace and automotive industries. While no one would argue that the science is mature and fully developed, the tools are available to gain significant insight into problems such as the vulnerability of aircraft. The models that to date have most fully combined these analyses concern nuclear weapons in abnormal environments (e.g., fires and crashes). The F-22 and JTCG/AS communities would be well advised to explore using the methodologies already perfected and still undergoing refinement in the U.S. nuclear weapon design laboratories.

The committee judges that a relatively large modeling uncertainty related to the F-22 is the inability to replicate the response of its new composite materials; this has already led to one surprise regarding the effects of hydraulic ram (see Chapter 4). While that particular problem has been fixed, there is always the possibility of more surprises.

Encounter Models

Almost by definition, encounter models are probabilistic. Initial conditions are quite variable, phenomena such as fragmentation are random, and the results are expressed as probabilities. While the events portrayed are probabilistic, the models usually calculate the probabilities in a deterministic way (i.e., by accepted probability rules). However, the Army is working with stochastic models (Deitz, 1995) toward achieving better interfacing with the results of live fire tests.

Models Used by the F-22 System Program Office

The F-22 SPO uses the following models in its work:

- FASTGEN 3 projects parallel rays through the target and describes intersections with aircraft components (Cramer and Hilbrand, 1985).
- FASTGEN 4 is similar to FASTGEN 3 but can use target information from a finite-element structural analysis model used in design. It was developed by members of the F-22 SPO (Griffis and Lentz, 1994).
- COVART 4.0 (Computation of Vulnerable Area and Repair Time) determines vulnerable areas for single kinetic-energy penetrators (or fragments) and high-explosive rounds. It was developed by the F-22 SPO (Bionetics Corporation, 1995). This extension of COVART 3.0 allows the incorporation of effects of small high-explosive rounds and includes special penetration equations for high-speed fragment impacts. Component defeat probabilities are calculated for each shot line and combined for target defeat probabilities for the individual fragment or round.
- The Blast Overpressure Analysis model is used to determine aircraft vulnerability to blast overpressure from conventional missile warheads. It is based on nuclear blast methodology that has been adapted to simulate the effects of smaller conventional warheads (Smith and Stewart, 1986).[2]
- The SHAZAM computer program is used to evaluate the effectiveness of an air-intercept missile by describing the terminal phase of the encounter. The program determines missile fuzing and detonation positions and calculates target damage sequentially from prioritized kill

---

[2] Regarding this model, the committee notes that the documentation it received was informal and of relatively poor quality, and the methodology is not used in the vulnerability community at large.

mechanisms (e.g., direct hit, blast overpressure, and fragment impacts) (Moore et al., 1994).
- The ESAMS (Enhanced Surface-to-Air Missile Simulator) simulates an encounter between the target and a radar-guided surface-to-Air missile. It provides a one-on-one framework in which to evaluate air vehicle survivability and optimization of tactics (BDM International, 1991).

With the possible exception of ESAMS, the committee understands that none of these models has been formally validated or accredited by either the JTCG/AS or the Joint Technical Coordinating Group on Munitions Effectiveness. Although the models represent the current state of the art of vulnerability analysis, it appears that much of their development has been accomplished by the services without the participation of the joint groups.

Large-Scale Effects

A major argument given for the need to perform tests on large assemblies or a full-up, full-scale aircraft is the inability of the vulnerability models to analyze adequately large-scale effects.[3] The encounter models merely reflect the capabilities of the phenomenological models, and the phenomenological models that are currently available tend to represent localized effects. This state of affairs reflects the empirical nature of these models and the impracticality of conducting an adequate experimental program to allow them to define completely large-scale damage.

As was discussed in Chapters 3 and 4, this committee believes that testing of a full-up, full-scale aircraft does not, at least in the case of a very expensive aircraft like the F-22, provide benefits judged worthy of the costs and will likely not yield useful data; however, testing at the subassembly level is worthwhile. Yet tests at the large subassembly level are more expensive than tests at the subsystem and component levels and less able to be repeated many times to yield statistically significant results.

Valid large-scale vulnerability models could provide an efficient means of making sound judgments about large-scale effects without the need for expensive and repetitive live fire tests on large subassemblies. Therefore, if models could be

---

[3] Large-scale effects could include the following: the effects of stress propagation in large portions of the target, interaction of damage with flight loads on the aircraft, propagation of fire and ignition, and response of loaded weapon bays to damage. Of course, very large effects that would simply destroy the aircraft do not need to be modeled at this level of detail.

developed that would lead to correct conclusions regarding large-scale effects, this committee would fully endorse that effort.

CONCLUSIONS

To be successful, vulnerability assessment requires much mutual support between documentation, data bases, models, and testing. This mutual support is as necessary for the F-22 program as for any other weapon system. The committee's review of the F-22 vulnerability assessment program indicated that significant improvements are needed in several of the tools that complement live fire testing.

With respect to documentation, most of the documents reviewed by the committee need to be updated and improved. Immediate attention to this matter by the DoD is warranted.

There is considerable need for expanded efforts over the next several years to improve the data bases used in models for conducting vulnerability assessments and planning live fire tests. Aggressive funding of joint live fire tests would enhance the data bases and could provide a major payoff for both the vulnerability reduction design and the live fire test of individual aircraft programs like the F-22.

Formal validation and accreditation, by the JTCG/AS and the Joint Technical Coordinating Group on Munitions Effectiveness, of models used by the Air Force and other services is warranted. The vulnerability community could make much better use of advanced analytical tools (e.g., finite-element analysis), especially in connection with understanding the response of F-22 composite materials to ballistic damage. Also, there is great merit to the development and exercise of numerical-analysis tools that will provide a better understanding of large-scale effects.

REFERENCES

BDM International. 1991. ESAMS Version 2.5 Users Manual, BDM/ABQ-89-0587-TR-R2. Albuquerque, N.M. April 12.

Bionetics Corporation. 1995. COVART 4.0 User Manual (Draft). Wright-Patterson Air Force Base, Ohio: F-22 System Program Office.

Cramer, R.E., and Hilbrand, R. 1985. FASTGEN 3 User's Manual. Wright-Patterson Air Force Base, Ohio: Aeronautical Systems Division, Air Force Systems Command.

Deitz, P. 1995. Army Vulnerability Testing—Results Methodology and Modeling. Presentation to the Committee on the Study of Live Fire Survivability Testing of the F-22 Aircraft, National Academy of Sciences, Washington, D.C., February 16.

DoD (U.S. Department of Defense). 1981a. Military Standard. Aircraft Nonnuclear Survivability Terms. MIL-STD-2089. Washington, D.C.: DoD.

DoD. 1981b. Military Standard. Requirements for Aircraft Nonnuclear Survivability Program. MIL-STD-2069. Washington, D.C.: DoD.

DoD. 1982. Military Handbook. Survivability, Aircraft, Nonnuclear, General Criteria—Volume 1. MIL-HDBK-336-1. Washington, D.C.: DoD.

DoD. 1983a. Military Handbook. Survivability, Aircraft, Nonnuclear, Airframe—Volume 2. MIL-HDBK-336-2. Washington, D.C.: DoD.

DoD. 1983b. Military Handbook. Survivability, Aircraft, Nonnuclear, Engine—Volume 3. MIL-HDBK-336-3. Washington, D.C.: DoD.

DoD. 1986. Military Standard. Survivability, Aeronautical Systems (For Combat Mission Effectiveness). MIL-STD-1799. Washington, D.C.: DoD.

Griffis, H., and Lentz, M. 1994. FASTGEN 4.1 User's Manual. XRESV Tech Note 94-01. Wright-Patterson Air Force Base, Ohio: Headquarters Aeronautical Systems Center.

Lauzze, R. 1995. Personal communication to Dale Atkinson, June 30.

Moore, C., D. Roberts, J. Martin, and H. Griffis. 1994. SHAZAM 2.0 User's Manual. XRESV Tech Note 94-02. Wright-Patterson Air Force Base, Ohio: Headquarters Aeronautical Systems Center.

Mowrer, D.W., R.G. Bernier, W. Enoch, R.E. Lake, and W.S. Vikestad. 1990. Aircraft Fuel System Fire and Explosion Suppression Design Guide. Aberdeen, Md.: Service Engineering Co.

NRC (National Research Council). 1993. Vulnerability Assessment of Aircraft: A Review of the Department of Defense Live Fire Test and Evaluation Program . Air Force Studies Board, NRC. Washington, D.C.: National Academy Press.

Smith, R.D., and W.M. Stewart. 1986. Documentation for the Blast Overpressure Analysis Model. Ft. Worth, Tex.: Lockheed Corp.

6

Recommendations

This chapter sets forth the committee's recommendations. These recommendations are based on the findings in the preceding chapters and directed at the specific matters in the legislation that requested this study (see Statement of Task in Preface).

The committee's principal recommendation, which appears immediately below, requires action by Congress. The numbered recommendations that follow require action by the DoD. Specific authorization and appropriation by Congress may be necessary to implement some of the numbered recommendations.

DESIRABILITY OF WAIVER FOR THE F-22 TESTS

Principal Recommendation. Permit a waiver of the full-up, full-scale, live fire tests required by law for the F-22. The committee believes that such tests are impractical and offer low benefits for the costs.

Recommendation 1. Interpret a waiver as reinforcing the need to conduct robust live fire tests of the F-22 that build incrementally from the component level to the subassembly or large assembly levels. (Recommendations to strengthen the current test program appear below.)

COST-BENEFIT METHODOLOGY

Recommendation 2. Continue DoD efforts to develop viable cost-benefit methodologies for planning the extent of live fire testing. Pursue methodologies to examine cost-benefit issues in the light of frameworks that take a broad view of how the future may develop for weapon systems like the F-22.

SUFFICIENCY OF TESTS PLANNED FOR THE F-22

The following recommendation applies to the replication of anti-air missile warhead threats against the F-22.

Recommendation 3. Consider, in future analyses and tests, the kill mechanism that involves dense multiple fragment impacts.

Although the Air Force and its contractors have constructed a comprehensive live fire test program, the following specific actions are recommended to strengthen the program as the F-22 proceeds with EMD and initial production.

Recommendation 4a. Conduct additional live fire testing to determine the damage that can be expected from a hit in the Frame 6 aft boom attachment area. Determine the most critical shot lines for this testing.

Recommendation 4b. Expand analyses to predict damage sizes and residual strengths of the aft boom, Frame 6, and horizontal tail pivot shafts after being hit by 30mm HEI rounds. Also, determine the risk of aircraft loss should it be found that loss of a horizontal tail is possible.

Recommendation 4c. Conduct further analysis of the aft fuel tank (A-1) prior to the conduct of Test 4D. Focus this analysis on determining the adequacy of the test specimen, with particular emphasis on its ability to simulate accurately the reaction of the entire tank.

Recommendation 4d. Make the operational community fully aware that a fuel ingestion risk to the aircraft exists at a fuel state higher than 60 percent. (This risk arises because the fuel tanks next to the engine inlets are not empty at fuel states above 60 percent; thus, a puncture could lead to fuel ingestion by an engine and potential engine failure.)

Recommendation 4e. Conduct the tests and analyses, proposed by the F-22 SPO, on the flammability of coolant and other fluids and the attendant vulnerability of the aircraft.

Recommendation 4f. Undertake the analysis and test, proposed by the SPO, of ablative materials in the weapons bay. Also, conduct further

RECOMMENDATIONS

analysis of the tradeoffs associated with additional ordnance protection or defensive measures.

Recommendation 4g. Fund the JTCG/AS and the Joint Live Fire Test Program to assure the completeness of data on the vulnerabilities of on-board ordnance.

Recommendation 4h. Fund the proposed Joint Live Fire testing of F119 engine components to alleviate the paucity of testing against those components.

Recommendation 4i. Emphasize continuing efforts by the F-22 SPO and JTCG/AS to develop improved methodologies for reducing flight crew vulnerability.

Recommendation 4j. Use the prototype air vehicle fuselage in Test 6A in lieu of a mock-up. (The Air Force has considered using this fuselage for Test 6A.)

In addition, the committee recommends that the Air Force begin planning for expeditious vulnerability assessment testing of the F-22 similar to that being conducted or planned for variants of the Navy's F-18.

Recommendation 5a. Use large subassemblies from production-representative hardware (e.g., a damaged aircraft or other source) in these tests.

Recommendation 5b. Provide these assets, as soon as they become available, to the vulnerability assessment community for the conduct of live fire tests.

Recommendation 5c. Direct the tests at (a) verifying predictions from the current F-22 live fire test program and the models used, and (b) testing the effects on overall F-22 vulnerability assessment brought about by configuration and mission changes. Also, use the tests to verify techniques for repairing battle damage to the F-22's new composite materials and systems.

OTHER RECOMMENDATIONS

Vulnerability Requirements

Recommendation 6. Reexamine expeditiously, for future F-22 missions (e.g., air-to-surface), the balance of requirements among susceptibility, vulnerability, and related performance parameters.

Recommendation 7. Include in operational requirements for any new missions user validation of quantitative vulnerability requirements, and plan new live fire tests as necessary in response to those requirements.

Vulnerability Assessment Tools

Recommendation 8. Update and improve expeditiously the various standards, handbooks, and design guides that are important to the aircraft vulnerability community.

Recommendation 9. Direct the JTCG/AS to define and plan a Joint Live Fire Test Program that will, over the next several years, produce sound vulnerability data bases; apply aggressive funding to implement this program.

Recommendation 10a. Validate and accredit formally, by the JTCG/AS and the Joint Technical Coordinating Group on Munitions Effectiveness, the vulnerability assessment models used by the Air Force and other services.

Recommendation 10b. Improve the vulnerability models of the vulnerability community, and adopt these improvements for the F-22.

Recommendation 10c. Explore the application of advanced methodologies currently being used by nuclear weapon designers and other industries.

Recommendation 10d. Focus on ways to understand fully the response of F-22 composite materials to ballistic damage, and develop and exercise analysis tools that can handle large-scale damage effects.

RECOMMENDATIONS

Appendix A

Meetings, Site Visits, and Discussions

COMMITTEE MEETING: DECEMBER 21-22, 1994 WASHINGTON, D.C.

Participants

Committee except Charles Crawford, Alan Epstein, Don Giadrosich, Robert Loewy (see page *iii* for a list of committee members); NRC staff (Mike Clarke, Bruce Braun, John Hughes, and Norm Haller); and briefers (listed with the presentations below).

Objectives

Complete administrative matters; agree on tasking and study plan; review tentative report outline; assign persons responsible for various sections of report; begin data gathering from selected presenters—Air Force, OSD, congressional staff, and others as appropriate; decide what additional data are needed; and determine next steps.

Presentations

Threat, Mission, and Operational Requirements for the F-22. Presented by Brig. Gen. William S. Hinton, Jr., U.S. Air Combat Command, Requirements.

OSD Views on Waiver. Presented by Dr. Albert Rainis, Office of the Secretary of Defense, Tactical Warfare Program.

Discussion of Waiver of Live Fire Tests. Presented by Lt. Gen. Richard E. Hawley, Office of the Secretary of the Air Force, Acquisition.

Overview of the F-22 Program. Presented by Maj. Gen. Robert F. Raggio, F-22 System Program Office.

APPENDIX A

Congressional Views on Waiver. Presented by Mark Forman, Senate Staff.

Discussion of Live Fire Testing Philosophy and the History Associated with First Report. Presented by James O'Bryon, Office of the Secretary of Defense, Live Fire Testing.

Air Force Test and Evaluation Plans for F-22 and Other Applicable Air Force Aircraft. Presented by Lt. Gen. Howard W. Leaf (Ret.), Air Force Test and Evaluation; Ralph Lauzze, Wright Laboratories; and Jon Ogg, F-22 System Program Office.

Discussion of Testing for Navy F-18. Presented by John Aldridge, Naval Air Systems Command.

Discussion of Other Views on Waivers in General. Presented by Louis J. Rodrigues, General Accounting Office.

COMMITTEE MEETING AND SITE VISIT: JANUARY 19-20, 1995 F-22 SYSTEM PROGRAM OFFICE, WRIGHT-PATTERSON AIR FORCE BASE, OHIO

Participants

Committee except Charles Crawford, Robert Loewy, and Larry Ullyatt; NRC staff; and briefers.

Objectives

Complete administrative matters; review updated report outline; continue data gathering from selected presenters representing OSD, the Army, the Air Force, and the Navy; further refine report storyboards; panel chairs brief report status; decide what additional data are needed; and determine next steps.

APPENDIX A 101

Presentations

Special Topics. Presented by LTC John Lawless, Joint Technical Coordinating Group; and Kevin Crosthwaite, Survivability/Vulnerability Information Analysis Center.

Vulnerability Program Overview. Presented by Jon Ogg, F-22 System Program Office.

Vulnerability Reduction Features. Presented by John Donnelly, Lockheed Corp., and Jim Shipman, Pratt and Whitney.

High Power Microwave. Presented by Joe Giorlando, Lockheed Corp.

Ballistic Vulnerability Analysis. Presented by Mark Stewart, Lockheed Corp., and Jim Shipman, Pratt and Whitney.

Live Fire Test Program. Presented by Hugh Griffis, F-22 System Program Office.

Joint Live Fire Testing. Presented by Ralph Lauzze, Wright Laboratories.

Cost Benefit Analysis Methodology. Presented by Hugh Griffis, F-22 System Program Office, and Ralph Lauzze, Wright Laboratories.

MEETING TO DISCUSS PREVIOUS NRC REPORT: FEBRUARY 8, 1995, SCIENCE APPLICATIONS INTERNATIONAL CORPORATION (SAIC), TYSONS CORNER, VIRGINIA

Participants

Committee member Harry Reed; NRC staff member Mike Clarke; Pete Adolph, SAIC (formerly with the Office of the Secretary of Defense); Albert Rainis, Office of the Secretary of Defense; and Larry Stanford, TRW.

APPENDIX A

Objective

Discuss responses to various findings in the previous NRC report regarding live fire testing, *Vulnerability Assessment of Aircraft*.

COMMITTEE MEETING: FEBRUARY 16-17, 1995 WASHINGTON, D.C.

Participants

Committee, NRC staff, briefers, and additional participants listed below.

Objectives

Continue data gathering through (a) briefings, and (b) a round-table question-and-answer session with invited participants.

Presentations

Live Fire Test and Evaluation of the F-22 Aircraft. Presented by Lowell Tonnessen and Larry Eusanio, Institute for Defense Analyses.

Navy Vulnerability Testing—Results, Methodology & Modeling. Presented by David Hall, Naval Air Warfare Center.

Army Vulnerability Testing—Results Methodology and Modeling. Presented by Paul H. Deitz, Army Research Laboratory.

Knowledge-Based Benefit/Cost Methodology for Live Fire Test Evaluation. Presented by Terry Klopcic, Army Research Laboratory.

Round-Table Question-and-Answer Session

Survivability and live fire testing issues were discussed by the committee, NRC staff, and the following participants:

Chuck Brammeier, Office of the Secretary of Defense, Test and Evaluation

Paul Deitz, Army Research Laboratory

Larry Eusanio, Institute for Defense Analyses

Lee Frame, Office of the Secretary of Defense, Test and Evaluation

Hugh Griffis, F-22 System Program Office

David Hall, Naval Air Warfare Center

LCDR David Hattery, Joint Technical Coordinating Group on Aircraft Survivability

Terry Klopcic, Army Research Laboratory

Ralph Lauzze, Wright Laboratories

LTC John Lawless, Joint Technical Coordinating Group on Aircraft Survivability

Jim O'Bryon, Office of the Secretary of Defense, Live Fire Testing

Al Rainis, Office of the Secretary of Defense

Arthur Stein, Institute for Defense Analyses

Jerry Wallick, Logistics Management Institute

SITE VISIT: FEBRUARY 21, 1995 NAVAL AIR WARFARE CENTER, CHINA LAKE, CALIFORNIA

Participants

Committee members Dale Atkinson, Charles Crawford, Alan Epstein, Donald Giadrosich, Robert Hillyet, and Milton Margolis; NRC staff members Mike Clarke and John Hughes; and representatives from the Air Force, the Institute for Defense Analyses, and the JTCG/AS.

Objective

To gather data regarding existing live fire test programs.

Presentations

Uses and Limits of Vulnerability Models. Presented by Dave Hall, Naval Air Warfare Center.

The Test Data Integration Process and Results. Presented by John Manion, Naval Air Warfare Center.

F/A-18E/F and V-22 Live Fire Test Program. Presented by J. Hardy Tyson and Tim Wise, Naval Air Warfare Center

APPENDIX A

COMMITTEE MEETING: MARCH 21-22, 1995 WASHINGTON, D.C.

Participants

Committee except Cynthia Volkerr and NRC staff.

Objective

Writing meeting held in executive session.

COMMITTEE MEETING: APRIL 27-28, 1995 WASHINGTON, D.C.

Participants

Committee except Larry Ullaytt; and NRC staff.

Objective

Writing meeting held in executive session.

PANEL MEETING: MARCH 3, 1995 ALBUQUERQUE, NEW MEXICO

Participants

Committee panel members John Bode, Delores Etter, Don Giadrosich, and Milton Margolis.

Objective

Writing meeting.

Appendix B

Live Fire Test Law U.S. Code, Title 10, Section 2366, 1994

2366. Major systems and munitions programs: survivability testing and lethality testing required before full-scale production

(a) Requirements—

(1) The Secretary of Defense shall provide that—

(A) a covered system may not proceed beyond low-rate initial production until realistic survivability testing of the system is completed in accordance with this section and the report required by subsection (d) with respect to that testing is submitted in accordance with that subsection; and

(B) a major munition program or a missile program may not proceed beyond low-rate initial production until realistic lethality testing of the program is completed in accordance with this section and the report required by subsection (d) with respect to that testing is submitted in accordance with that subsection.

(2) The Secretary of Defense shall provide that a covered product improvement program may not proceed beyond low-rate initial production until—

(A) in the case of a product improvement to a covered system, realistic survivability testing is completed in accordance with this section; and

(B) in the case of a product improvement to a major munitions program or a missile program, realistic lethality testing is completed in accordance with this section.

(b) Test guidelines—

(1) Survivability and lethality tests required under subsection (a) shall be carried out sufficiently early in the development phase of the system or program (including a covered product improvement program) to allow any design deficiency demonstrated by the testing to be corrected in the design of the system, munition, or missile (or in the product modification or upgrade to the system, munition, or missile) before proceeding beyond low-rate initial production.

(2) The costs of all tests required under that subsection shall be paid from funds available for the system being tested.

(c) Waiver authority—

(1) The Secretary of Defense may waive the application of the survivability and lethality tests of this section to a covered system, munitions program, missile program, or covered product improvement program if the Secretary, before the system or program enters engineering and manufacturing development, certifies to Congress that live-fire testing of such system or program would be unreasonably expensive and impractical.

(2) In the case of a covered system (or covered product improvement program for a covered system), the Secretary may waive the application of the survivability and lethality tests of this section to such system or program and instead allow testing of the system or program in combat by firing munitions likely to be encountered in combat at components, subsystems, and subassemblies, together with performing design analyses, modeling and simulation, and analysis of combat data. Such alternative testing may not be carried out in the case of any covered system (or covered product improvement program for a covered system) unless the Secretary certifies to Congress, before the system or program enters engineering and manufacturing development, that the survivability and lethality testing of such system or program otherwise required by this section would be unreasonably expensive and impracticable.

(3) The Secretary shall include with any certification under paragraph (1) or (2) a report explaining how the Secretary plans to evaluate the survivability or the lethality of the system or program and assessing possible alternatives to realistic survivability testing of the system or program.

(4) In time of war or mobilization, the President may suspend the operation of any provision of this section.

(d) Reporting to Congress—At the conclusion of survivability or lethality testing under subsection (a), the Secretary of Defense shall submit a report on the testing to the Committees on Armed Services and on Appropriations of the Senate and House of Representatives. Each such report shall describe the results of the survivability or lethality testing and shall give the Secretary's overall assessment of the testing.

(e) Definitions—In this section:

(1) The term "covered system" means a vehicle, weapon platform, or conventional weapon system—

(A) that includes features designed to provide some degree of protection to users in combat; and

(B) that is a major system within the meaning of that term in section 2302(5) of this title.

(2) The term "major munitions program" means—

(A) a munition program for which more than 1,000,000 rounds are planned to be acquired; or

(B) a conventional munitions program that is a major system within the meaning of that term in section 2302(5) of this title.

(3) The term "realistic survivability testing" means, in the case of a covered system (or a covered product improvement program for a covered system), testing for vulnerability of the system in combat by firing munitions likely to be encountered in combat (or munitions with a capability similar to such munitions) at the system configured for combat, with the primary emphasis on testing vulnerability with respect to potential user casualties and taking into equal consideration the susceptibility to attack and combat performance of the system.

(4) The term "realistic lethality testing" means, in the case of a major munitions program or a missile program (or a covered product improvement program for such a program), testing for lethality by firing the munition or missile concerned at appropriate targets configured for combat.

(5) The term "configured for combat," with respect to a weapon system, platform, or vehicle, means loaded or equipped with all dangerous materials (including all flammables and explosives) that would normally be on board in combat.

(6) The term "covered product improvement program" means a program under which—

(A) a modification or upgrade will be made to a covered system which (as determined by the Secretary of Defense) is likely to affect significantly the survivability of such system; or

(B) a modification or upgrade will be made to a major munitions program or a missile program which (as determined by the Secretary of Defense) is likely to affect significantly the lethality of the munition or missile produced under the program.

Appendix C

Department of Defense F-22 Waiver Request

This appendix reproduces verbatim the letter submitted by the Department of Defense to Congress (Letterhead of the General Counsel of the Department of Defense, Washington, D.C. 20301-1600, dated October 8, 1993) requesting a waiver of live fire testing for the F-22. Included are attachments (a) Draft Legislation; (b) Plan for Alternative Assessment; and (c) Section by Section Analysis.

APPENDIX C

The Honorable Al Gore
President of the Senate
Washington, DC 20510

Dear Mr. President:

Enclosed is draft legislation, "To authorize a retroactive waiver of the survivability and lethality testing procedures that apply to the F-22 program."

This proposal is part of the Department of Defense Legislative Program for the 103d Congress and the Office of Management and Budget advises that, from the standpoint of the Administration's program, there is no objection to the presentation of this proposal for the consideration of Congress.

<u>Purpose of the Legislation</u>

Section 2366 of title 10, United States Code, requires realistic survivability and lethality testing of covered systems and munitions programs prior to full-rate production. The requirement is that the covered system must be tested for vulnerability in combat by firing munitions, likely to be encountered in combat, at the system configured for combat.

Section 2366 of title 10 allows the Secretary of Defense to waive the requirement if, before the system enters full-scale engineering development, the Secretary certifies to Congress that live fire testing of the system would be unreasonably expensive and impractical. Because of the cost of an F-22 aircraft, such testing is both unreasonably expensive and impractical. Since the F-22 has already entered full-scale engineering development, legislation is needed to allow the Secretary of Defense to grant a waiver.

In order for the Secretary of Defense to evaluate the survivability of the F-22 aircraft, the Air Force developed the revised live fire test program that is summarized in an enclosure to this letter. This plan includes detailed analyses, review of historical test data, and incremental build-up testing that includes material characterization tests and live fire testing of selected components and subassemblies. Information from the results of these tests will be taken into account in the F-22's design. In this way, we plan to achieve fully the objective of section 2366 in as realistic a manner as is consistent with cost effectiveness and practicality.

The proposed legislation will authorize the Secretary of Defense to grant a waiver to the survivability testing requirements in section 2366 as they apply to the F-22 system.

<u>Cost and Budget Data</u>

The enactment of this legislative proposal shall not cause any increase in appropriated funding for the Department of Defense or have any budgetary impact.

Sincerely,
[Signed Jamie S. Gorelick]
Jamie S. Gorelick

Enclosures:
Draft Legislation
Plan for Alternative Assessment
Section by Section Analysis

A BILL [ENCLOSURE A]

To authorize a retroactive waiver of the survivability testing procedures that apply to the F-22 program

<u>Be it enacted by the Senate and House of Representatives of the United States of America, in Congress assembled</u>, That

Section Live-Fire Survivability Testing of F-22 Aircraft

(a) **Authority for Retroactive Waiver.** —The Secretary of Defense may exercise the waiver authority in section 2366(c) of title 10, United States Code, with respect to the application of the survivability tests of that section to the F-22 aircraft, notwithstanding that such program has entered full-scale engineering development.

(b) **Reporting Requirement.** —If the Secretary of Defense submits a certification under section 2366(c) of such title 10 that live-fire testing of the F-22 system under such section would be unreasonably expensive or impractical, the Secretary of Defense shall require that sufficiently large and realistic components and subsystems that could affect the survivability of the F-22 system be made available for any alternative live-fire test program.

(c) **Funding.** —The funds required to carry out any alternative live-fire testing program for the F-22 aircraft system shall be made available from amounts appropriated for the F-22 program.

APPENDIX C

PLAN FOR ALTERNATIVE ASSESSMENT OF THE VULNERABILITY OF THE F-22 AIRCRAFT [ENCLOSURE B]

Executive Summary

The Air Force, in consultation with the Office of the Secretary of Defense, proposes the following plan to evaluate the survivability of the F-22 aircraft. In order to assess the adequacy of the testing and to provide the report to Congress required by section 2366 of title 10, United States Code, the Office of the Secretary of Defense will review and comment to the Air Force on their test and evaluation plans, will observe testing, will obtain all relevant test results from the Air Force in a timely manner, and will review the Air Force's Live Fire Test and Evaluation (LFT&E) report.

The Air Force is conducting a planned vulnerability reduction program which includes the LFT&E program. The F-22 vulnerability reduction/LFT&E program uses detailed requirements analyses, vulnerability reduction design features, and a build-up ballistic testing approach to verify the combat survivability of the air vehicle design. Tests are being completed in parallel with the Engineering and Manufacturing Development (ENO) [sic] design activity. This concurrent design/test approach ensures vulnerability reduction features are addressed as early as possible in the development process, thus resulting in overall cost and risk reduction.

The data we collected by combining the F-22 Demonstration and Validation (Dem/Val) Program with the EMD vulnerability reduction program showed two major damage mechanisms that are potentially the most serious and likely sources of aircraft loss:

- Fire/explosion within the dry bay and/or fuel tank ullage areas.
- Hydrodynamic ram-induced structural failure of the fuel cells.

The Air Force assessment plan calls for extensive analysis of these potential damage mechanisms, followed by testing of unique materials, components and then larger realistic aircraft subsystems to verify the predicted vulnerability results.

Several other vulnerability issues have been addressed in the Air Force plan. These are:

- Separation and redundancy of critical components, hydraulic and electrical lines.
- Flight and engine control.
- On-board ordnance.
- Directed energy threats.

APPENDIX C

- Chemical threats.
- Crew casualties.

Some of these issues currently are being addressed by analysis, testing, or, in most cases, a combination of the two. The current LFT plan calls for continuing analysis of these issues.

Systems Engineering Approach

The vulnerability reduction program is based upon the system engineering process of understanding user needs, defining vulnerability requirements which meet system requirements, developing balanced design solutions and testing. Each of these pieces of the program will be described in the following paragraphs.

Requirements Definition.

The baseline threats are derived from the Operational Requirements Document (ORD) and documented in the System Threat Assessment Report (STAR). Given these threats, the results of vulnerability analysis are compared against system requirements. An iterative design process continues until the vulnerability level meets the system requirements. As an example, in the Dem/Val program, three major iterations occurred before the vulnerability levels were determined to be adequate.

Detailed Analyses.

The vulnerability analysis is based upon standard digital computer models, such as those sponsored by the Joint Technical Coordinating Group/Aircraft Survivability (JTCG/AS). Major elements of the vulnerability assessment include the threat definition, Failure Mode Effects and Criticality Analyses (FMECA), Damage Modes and Effects Analysis (DMEA), system description, shotline generation model, and component probability of kill given a hit (Pk/h) curves. The analysis outputs are vulnerable areas and overall system Pk/h.

Ballistic Tests.

Ballistic tests will be conducted to fill voids in historical vulnerability data, obtain test data on unique F-22 materials or components, and to verify model analyses. During the EMD Program both development and verification testing are considered part of the F-22 vulnerability reduction/LFT program. Some of the planned major tests are outlined below.

Test Descriptions

Three categories of ballistic tests will be performed: 1) material, 2) component, and 3) subassembly.

APPENDIX C

Material Tests.

Penetration equations are required for vulnerability analysis. These equations are used to predict whether a specified projectile or warhead fragment can penetrate the aircraft materials it impacts along its trajectory and predict the residual mass and velocity after impact. Penetration equations for a material unique to the F-22 do not exist. Penetration tests will be performed on small panels of this material. Projectile type and size, impact angle, impact velocity, and material thickness will be test variables. Test results will be used to develop penetration equations and fire starting equations.

Component Tests.

Component probability of kill given a hit (Pk/h) data are the heart of any vulnerability analysis. Component Pk/h curves for all critical components will be determined by analysis. A selected sample of components will be tested to verify the analysis method used to calculate the component Pk/h data. The selection of components to be tested will be based upon component availability, critical nature of the component, and uncertainty in the component Pk/h analysis.

Subassembly Tests.

A number of subassembly tests will be conducted to cover the critical regions of the aircraft as described below.

Wing Box Subassembly Tests.

Hydrodynamic ram phenomena have not been adequately modeled. Hydrodynamic ram testing comprises a significant portion of the F-22 ballistic testing because of its importance in aircraft survivability and because of model inadequacies. Tests to determine hydrodynamic ram damage to the wings will start with a small section of wing (wing box) containing 3 spars and eventually build up in a step-wise manner to larger sections containing 8 spars. The first tests will be used to identify design features to withstand hydrodynamic ram effects.

The final test will be used to verify that the design finally selected can survive the specified threats. (A 30mm HEI round will be used in this test.) It will demonstrate the survivability of the wing design to both hydrodynamic ram effects and fire/explosion in the ullage (air space in the fuel tank above the fuel) under expected combat conditions. Fuel tank ullage will be inerted to simulate the effects of the On-Board Inert Gas Generation System (OBIGGS). During the test, the test article will be subjected to an airflow and will be structurally loaded to simulate flight conditions.

The conditions established for the above test assume that the OBIGGS has adequate capacity to produce sufficient inertant for all expected flight profiles and has a distribution system to adequately disseminate the inertant gas within all of the fuel tanks. The capabilities of the OBIGGS design to produce and distribute adequate inertant will be verified by analysis and tests.

Aft Side of Body Subassembly Tests.

Two aft side of body (fuselage) tests are planned. The first test will demonstrate that a 30mm HEI round will not cause massive damage to this structure.

Depending upon the outcome of the first test and an updated vulnerability assessment, a second test may be conducted. This test article would be constructed using portions of the Prototype Air Vehicle (PAV), upgraded to represent the current F-22 design. The test article would be subjected to airflow and would serve to evaluate both synergistic effects and separation/isolation schemes.

Fuselage Fuel Tank Subassembly.

The fuselage fuel tank structural test article will include the aft crew station bulkhead. This development test will demonstrate that structural (hydrodynamic ram) damage does not result in a catastrophic (structural or air crew) failure. The air crew is shielded by the aft crew station bulkhead. The tank will be filled with water to the combat fuel load level. The threat will be a 30mm HEI or API projectile.

Dry Bay Subassembly Tests.

The primary objective of this test series is to address the uncertainties associated with synergistic effects; i.e., the interaction of damaged components on nearby components. A test article will be fabricated to represent various F-22 configurations and will include F-22 representative components. The tests will be conducted to demonstrate the synergistic effects of electrical, hydraulic, fuel, and coolant components. Actual fluids (fuel, hydraulic, coolant) will be used. Airflow will be provided as required to simulate internal and external conditions. Various combinations of threats and shotlines will be tested.

In addition, fire protection system effectiveness will be demonstrated. Some of the tests in the series will include active fire suppression systems identical to those used on the F-22.

Engine Tests.

The results of the extensive ballistic testing on the F-15's F 100 engine are applicable to the F-22's F119 engine. The response of the engine diffuser case and the Full Authority Digital Engine Control (FADEC) to ballistic threats are of particular interest. Test results which apply to the F-22 are being factored into the vulnerability analysis.

The F-22's F-119 engine introduces thrust vectoring to combat aircraft. An analysis will be conducted to determine whether a ballistic impact can cause the first vectoring system to lock in a hard-over position. If it appears that ballistic impacts or Directed Energy Weapons can produce uncontrollable flight conditions, then the vulnerability analysis will be modified to reflect this damage mode. If needed, testing will be accomplished to confirm this analysis.

On-Board Ordnance.

The F-22 SPO is monitoring the on-board ordnance testing being conducted as part of the Joint Live Fire (JLF) Program. F-22 SPO engineers have attended meetings with Wright Laboratory test engineers leading the test planning effort, so this series of tests can be performed in a manner which will provide information useful to the F-22 program. Test results which can be applied to the F-22 will be factored into the F-22 vulnerability program. Depending on the review of the generic test results, some F-22 specific testing may be required to confirm the vulnerability analysis.

Aircraft Battle Damage Repair (ABDR).

In conjunction with subassembly ballistic testing, ABDR procedures and techniques will be developed, validated, and verified. The Air Force ABDR Program Office from Sacramento Air Logistics Center will participate in this activity. The above ballistic test articles will be made available for ABDR activities.

High Power Microwave (HPM) Test.

HPM tests will include coupling energy into wires and connectors as well as conducted and radiated antenna tests.

Laser Tests.

Laser testing will be conducted on applicable components based upon vulnerability reduction requirements.

Chemical Testing.

The F-22 weapon system is being hardened to withstand chemical weapons. The effects of chemical agents on the F-22 materials is being tested using coupons (small panels) of the materials in question. Based upon this coupon testing, materials and coatings will be selected. Selection of chemical resistant materials and coatings will improve the F-22's ability to operate in a chemical environment.

Hardening a fighter aircraft to chemical weapons has never been attempted. Hence, the F-22 program has requested and received support from the Human Systems Center (HSC) at Brooks Air Force Base and U.S. Army Dugway Proving Grounds (DPG). HSC and DPG plan to perform a series of system level "proof-of-concept" decontamination tests on a surrogate fighter aircraft. Availability of this test data will allow F-22 designers to make timely and informed design decisions. Late in EMD, the F-22 air vehicle will be exposed to a chemical agent simulant and then be decontaminated to demonstrate the F-22 decontamination capability.

Model Enhancements

As stated earlier, some of the ballistic tests selected were based upon the fact that existing models are not adequate for all situations of interest. For example, they are not adequate for predicting penetration of new materials, for predicting sustained fires, and for predicting damage when there are synergistic effects. Ballistic test results will be used to reduce the uncertainties. Enhanced models will in turn provide increased fidelity and confidence in the vulnerability analyses. The

F-22 SPO has initiated vulnerability model code enhancements. Additionally, the F-22 SPO has requested and received Joint Technical Coordinating Group (JTCG) support in accomplishing additional improvements.

Crew Casualties

Crew casualty reduction is one of the critical factors identified in the Live Fire Test legislation. A number of the tests described above will yield information which will be used to minimize crew casualties. In keeping with the spirit of the legislation, an effort has been made to design the aircraft for reduced casualties from all sources (e.g., ballistic impact, fire, smoke) and to facilitate the safe escape of the crew in the event the aircraft is lost.

SECTION BY SECTION ANALYSIS [ENCLOSURE C]

Live-Fire Survivability Testing of F-22 Aircraft

This amendment would require the Secretary of Defense to submit a report explaining how the Secretary plans to evaluate the survivability of the F-22 system and assessing various alternatives to realistic survivability testing. The provision also would require the Secretary to ensure that major components and subsystems that could significantly affect the survivability of the F-22 be made available for live-fire testing.

Appendix D

Vulnerability Assessment Process

The body of this appendix reproduces pages 11 to 18 (numbered here as pages D-1 to D-8) of the National Research Council's 1993 report *Vulnerability Assessment of Aircraft*. Citations called out within the text are included in a reference section from the original report, at the end of the text.

WHAT ARE THE THREATS TO MILITARY AIRCRAFT?

When the military began to use aircraft in war, the opposing forces began using weapons in an attempt to destroy them. In the first half of the twentieth century, guns were the primary weapons used against aircraft. These guns were either surface-based or carried by enemy aircraft. They ranged from the small arms weapons, such as the 0.3/0.303-inch (7.62/7.7-millimeter) and 0.50-caliber (12.7-millimeter) machine guns, to anti-aircraft artillery (AAA), such as the 40-millimeter and 88-millimeter caliber guns of World War II (WW II). Contemporary guns that can be used against aircraft include the 5.56-millimeter, 7.62-millimeter, 12.7-millimeter, 14.5-millimeter, and 20-millimeter small arms, and the 23-millimeter, 30-millimeter, 37-millimeter, 57-millimeter, 76-millimeter, 85-millimeter, and 120-millimeter AAA. The small arms weapons typically fire ball ammunition, or armor-piercing projectiles, known as AP rounds, or AP projectiles with incendiaries, known as API rounds. The AAA weapons and the larger-caliber aircraft guns usually fire ballistic projectiles with a high-explosive (HE) core and a surrounding metal case. These are referred to as HE warheads or HEI warheads when incendiaries are included.[2] The HE warheads may detonate on contact with the aircraft (contact-fuzed HE warheads), after an elapsed time since firing (time-fuzed HE warheads), or in proximity to the aircraft (proximity-fuzed HE warheads).

After World War II, guided missiles, both surface-based and airborne, were developed to kill aircraft. These antiair weapons typically carry contact- or proximity-fuzed HE warheads designed to kill aircraft with fragments and blast. Guns and guided missiles are still the primary threat faced by aircraft today. However, several new threats to aircraft are in development. Directed energy weapons, in the form of low-to-medium power lasers and high-power microwaves, have the potential to damage or destroy sensors on the aircraft and the weapons they are carrying; and high-power lasers can damage major aircraft structure. Chemical and biological weapons pose a threat to aircraft, particularly on the surface, and nuclear weapons are a threat to aircraft on the surface and in the air.

WHAT IS AIRCRAFT VULNERABILITY?

Aircraft survive a mission into hostile territory by "avoiding" the damage-causing mechanisms of the enemy's air defense and by "withstanding" the damage caused by these mechanisms when they cannot be avoided. The aircraft attribute known as susceptibility refers to the inability of

---

[1] Much of the material presented in this chapter is based upon Ball (1985).
[2] Some of the small-caliber AAA also fire API rounds.

the aircraft to avoid (being damaged by) the man-made hostile environment and is measured by P_H, the probability the aircraft is hit by a weapon while on its mission. The aircraft attribute known as vulnerability refers to the inability of the aircraft to withstand (the damage caused by the) hostile environment and is measured by $P_{K/H}$, the probability the aircraft is killed[3] given that it is hit. The probability the aircraft is killed by a particular weapon while on the mission is P_K, which is equal to $P_H \cdot P_{K/H}$. The probability the aircraft survives the encounter with the weapon is P_S, which is equal to $1 - P_K$, which is the same as $1 - P_H \cdot P_{K/H}$. Thus, reducing an aircraft's susceptibility (P_H) and vulnerability ($P_{K/H}$) to the weapons likely to be encountered in combat increases its survivability. An aircraft's susceptibility can be reduced by destroying the enemy air defense elements, by reducing the aircraft's signatures (stealth), by employing on-board and off-board threat warning systems and electronic countermeasures, and by the tactics employed. An aircraft's vulnerability can be reduced by using redundant and separated components, by locating components to minimize the possibility and extent of damage, by designing components to contain or withstand the effects of damage, by adding special equipment to suppress the damage, by shielding components, and by removing vulnerable components from the design. A very important aspect of vulnerability reduction is that many design features are effective against a number of different threat weapons. For example, locating redundant flight control hydraulic components on opposite sides of the aircraft and inerting the fuel tank ullages will provide protection from both gun projectiles and proximity-fuzed missiles in most situations. Thus, in many situations it is not necessary to consider all of the individual threats when designing the aircraft.

Critical Components and Essential Functions.

Each component in the aircraft has a level, degree, or amount of vulnerability to the damage-causing mechanisms[4] generated by the threat weapon; and each component's vulnerability contributes in some measure to the vulnerability of the total aircraft. The critical components on an aircraft are those components whose kill result in the loss of an essential function. Essential functions are those functions required to prevent an aircraft kill. The essential functions that prevent an attrition kill are lift, thrust, and control of flight, and the ability to land safely. Navigation and weapons delivery are two possible essential functions for a mission abort kill. An example of a critical component for the attrition kill is the single pilot who controls the flight of the aircraft. If the pilot is killed (i.e., he/she is unable to perform the essential function of control of the aircraft) the aircraft is also killed. An example of a critical component on an attack aircraft for the mission abort kill is the weapons delivery computer. If the computer is killed, the weapons cannot be released at the correct time; consequently, the pilot will return to base prior to mission completion.

Components that do not contribute to any of the essential functions become critical when their response to a hit (i.e., their kill mode) causes the kill of another component that is critical because it contributes to an essential function. For example, consider the bombs carried on-board an attack aircraft. The bombs do not contribute to the essential functions for flight of lift, thrust, and control. However, if one of the bombs explodes when hit by a fragment or bullet, and the explosion kills the pilot or any other critical components on the aircraft, the bombs are critical components because their kill mode (explosion) eventually leads to a kill of the aircraft.[5] The propagation of damage from the hit component to other components is known as cascading damage. Pyrotechnic items, such as infrared flares, are also critical components when their reaction to a hit leads to a fire and the eventual loss of the aircraft.

The critical components can be nonredundant, such as the single pilot and single engine on a single-piloted, single-engined aircraft, or redundant, such as the two engines on a two-engined aircraft. When the critical components are redundant, a kill of more than one of the redundant components is required for a kill of the aircraft. In general, the critical components on a particular aircraft depend only upon the selected kill category (and level, if appropriate) and the assumed kill mode(s), and not upon the threat weapon.[6]

The procedure used to determine all of the nonredundant and redundant critical components on an aircraft is known as the critical component analysis. Two different types of analyses can be used, the Failure Mode and Effects Analysis (FMEA) and the Fault Tree Analysis (FTA). In the FMEA, all possible failure, damage, or kill modes of a component or subsystem are identified and the consequence of each

---

[3] The word kill is used here in a general sense. The vulnerability assessment community uses several definitions of kill. Two categories of kill are the attrition kill and the mission abort kill. There are several levels of attrition kill based upon the elapsed time of kill after the hit. For example, the K-level attrition kill is defined as a kill in which the aircraft falls out of control within 30 seconds after the hit, and the A level is defined as a kill in which the aircraft falls out of control within 5 minutes after the hit.

[4] Damage, threat, or kill mechanisms are the output of the threat warhead that cause damage to the aircraft. The types of damage mechanisms associated with penetrator and high-explosive warheads are penetrators, fragments, incendiaries, and blast. Damage processes refer to the interaction of the damage mechanism with the aircraft and its components. The damage processes associated with the damage mechanisms listed here include ballistic impact, penetration, combustion (in the form of a fire or explosion), hydraulic or hydrodynamic ram, and blast loading.

[5] The treatment of the on-board munitions when assessing aircraft vulnerability is a major concern to the committee, particularly for aircraft with internal ordnance storage. This concern is examined in detail in Chapters 2 and 4.

[6] Refer to footnote 3 for several examples of kill definitions.

TABLE 1-1 List of Some Subsystem Damage-Caused Failure (Kill) Modes [Ball, 1985]

Fuel Subsystem	*Propulsion Subsystem*	*Flight Control Subsystem*
Fuel supply depletion In-tank fire/ explosion Void space fire/explosion Sustained exterior fire Hydraulic ram	Fuel ingestion Foreign object ingestion Inlet flow distortion Lubrication starvation Compressor case perforation Combustor case perforation Turbine section failure	Disruption of control path Loss of control power Loss of aircraft motion data Damage to control surfaces Hydraulic fluid fire
Power Train/Rotor Blade/Propellor Subsystem Loss of lubrication Mechanical/ structural damage	Exhaust duct failure Engine control/ accessories failure	*Structural Subsystem* Structural removal Pressure overload
Electrical Subsystem Severing or grounding Mechanical failure Overheating	*Crew Subsystem* Injury, incapacitation, or death	Thermal weakening Penetration
	Armament Subsystem Fire/explosion	*Avionics Subsystem* Penetrator/fragment damage Fire/ explosion/overheat

component failure/damage/kill mode upon each of the essential functions is determined.[7] In the FTA, those component or subsystem kill modes required to cause the loss of the essential functions are determined.

Kill Modes.

For many years, the aircraft vulnerability community has observed the results of live fire testing of components, subsystems, and aircraft and has examined the combat data on damaged and killed aircraft in order to determine all of the kill modes associated with each of the aircraft subsystems. For example, there are five kill modes associated with the fuel subsystem. When a fuel tank is holed by a penetrator or fragment, a catastrophic explosion or major fire may occur inside the tank, or fuel may leak from the hole in the tank into an adjacent void space or dry bay and catch fire, or hydraulic ram damage to the fuel tank wall may cause a major structural failure of the tank or allow fuel to dump into engine intake ducts, causing an engine kill. A list of some of the possible kill modes for each of the major subsystems on an aircraft has been compiled based upon these observations and studies. This list is presented in Table 1-1.

The kill modes listed in Table 1-1 describe different types of reaction that components or subsystems in the aircraft exhibit when the aircraft is hit. In some of the kill modes, the component hit is the only component killed, whereas in others, the component hit reacts to the hit in a mode that kills other components. An example of the former is the loss of flight control due to a hit in a hydraulic power actuator that causes a jam of the actuator and a loss of control of the control surface. An example of the latter is a fuel ingestion kill of an engine due to a hit on a fuel tank adjacent to the air inlet. Reducing the vulnerability of an aircraft to the threat weapons and their damage mechanisms involves reducing the likelihood the kill modes given in Table 1-1 will occur when the aircraft is hit.

The Failure Mode and Effects Analysis (FMEA).

As an example of the FMEA process, consider a single-engine aircraft with only two fuel tanks, one in each wing. The tanks are partially full, and there are fuel vapors in the ullage[8] of the tanks. The possible kill modes for the fuel subsystem are given in Table 1-1. One fuel tank kill mode is an explosion inside the tank. If the consequence of the internal explosion in either wing tank is the destruction of the wing containing the tank, which then causes a kill of the aircraft due to loss of lift, both wing fuel tanks are nonredundant critical components for the attrition kill for the internal explosion kill mode. On the other hand, suppose the kill mode of the tanks is a loss of fuel storage capability due to one or more holes in the bottom of the tank. If this kill mode occurs in only one tank, this will not lead to a loss of thrust due to fuel supply depletion when the undamaged tank can provide fuel to the engine. However, if both tanks are holed and lose their storage capability, then a fuel supply depletion

---

[7] The relation between a component or system failure mode and combat-caused damage or kill modes is developed in the Damage Mode and Effects Analysis.

[8] The ullage is the volume of the tank above the fuel level. Fuel vapors accumulate in the ullage.

kill will occur, the aircraft will lose thrust, and an attrition kill will result. Thus, for this kill mode, the fuel tanks are redundant critical components.

The Fault Tree Analysis (FTA).

In the FTA process, the selected kill category (and possibly level) is defined as the top-level undesirable event, and the component kill required to cause the undesirable event are determined. The component kill that result in the undesired event are linked together in the fault tree by using logical AND and OR gates. For example, consider an aircraft with components A, B, C, and D. An undesirable kill will occur if either component A OR B is killed, or it may occur if both components C AND D are killed. Thus, components A and B are nonredundant critical components, and components C and D are redundant critical components. In using FTA for the fuel tank example given above, one undesirable event leading to an attrition kill is loss of lift. If loss of lift occurs due to an explosion inside the left wing fuel tank, a component A kill, OR if it occurs due to an explosion inside the right wing tank, a component B kill, both wing fuel tanks are nonredundant critical components for the explosion kill mode. On the other hand, a loss of thrust will occur if wing tanks A AND B are killed (by the fuel supply depletion kill mode). Thus, the tanks are redundant critical components for this kill mode. As another example of FTA, consider a two-engined aircraft. The undesired event of loss of thrust, which leads to an attrition kill, will occur when the left engine AND the right engine are killed. Thus, these two components are redundant critical components. A list of the typical critical components on a single-piloted, two-engined helicopter is given in Table 1-2.

TABLE 1-2 List of Typical Nonredundant and Redundant Critical Components on a Single-Piloted, Two-Engined Helicopter (Ball, 1985)

Nonredundant Critical Components	Redundant Critical Components
Flight Control Subsystem Components Rods, bellcranks, pitch links, swashplate, hydraulic actuators, collective lever, and control pedals	*Propulsion Subsystem Components* Engines and engine mounts
	Hydraulic Subsystem Components Hydraulic reservoirs, lines, and components
Rotor Blade and Power Train Components Blades, drive shafts, rotor heads, main transmission, and gearboxes	
	Structural Subsystem Components Redundant structural elements
Fuel Subsystem Components Fuel cells, sump, lines, and valves *Structural Subsystem Components* Tail boom	

The Kill Tree.

A visual illustration of all of the critical components and their redundancies is provided by the kill tree,[9] such as the one shown in Figure 1-1 for an attrition kill of a two-engined, two-piloted helicopter. A complete horizontal or diagonal cut through the tree trunk anywhere along the trunk will cause a kill. For example, a kill of the pilot and either the copilot or the copilot's controls will cause a kill, as will a kill of the drive train or any of the three cyclic actuators. If the kill mode of the left- and right-hand fuel tanks is fuel supply depletion, both tanks must be killed to cause a kill of the aircraft. On the other hand, if the kill mode is a fuel fire or explosion, then a kill of either tank will kill the aircraft. Once the critical components have been identified and arranged in the kill tree, a vulnerability assessment can be performed.

WHAT IS A VULNERABILITY ASSESSMENT?

A vulnerability assessment is broadly defined here as the systematic description, delineation, test and evaluation, analysis, or quantification of the vulnerability of the individual critical components and of the total aircraft. When an aircraft is hit by one or more damage mechanisms generated by the threat weapon, the outcome of those hits is not deterministic; it is random or stochastic.[10] For example, when 15 fragments from a proximity-fuzed high-explosive warhead penetrate the upper wall of an aircraft's wing fuel tank, the flammable vapor inside the tank may explode, destroying the wing and killing the aircraft; or the vapor may not

---

[9] The kill tree is also referred to as the fault tree.

[10] A deterministic process has a repeatable outcome that can be predicted with certainty if all of the influencing parameters and governing laws are known. Random or stochastic processes have multiple or various outcomes, any one of which may or may not occur on any one trial.

APPENDIX D

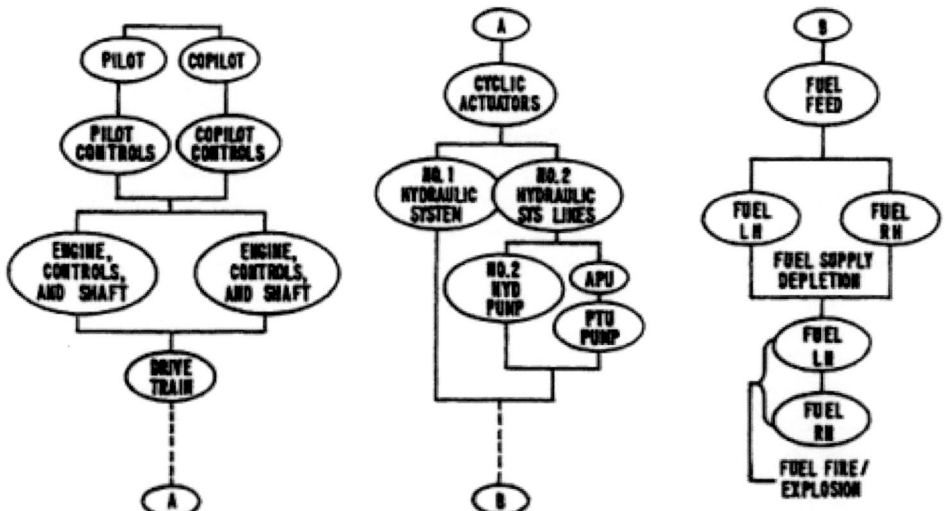

Figure 1-1 The attrition kill tree for a two-piloted, two-engined helicopter (Ball, 1985). Copyright © AIAA 1985— Used with permission.

explode, and the aircraft survives the 15 hits. The likelihood of an explosion inside the tank depends upon many random variables, such as the amount of fuel vapor, the oxygen concentration in the vicinity of the fragments, and the temperature of the fragments.

HOW IS VULNERABILITY MEASURED?

As a consequence of the random nature of vulnerability, the metric most often used to quantify the vulnerability of an aircraft's critical components is $P_{k/h}$, the probability the component is killed given a random hit on the component by a threat weapon or damage mechanism.[11] The value of $P_{k/h}$ depends upon the intensity of the terminal effects parameters associated with the damage mechanism, such as mass and impact velocity on the component for penetrators and fragments. The set of component $P_{k/h}$ values for different masses and impact velocities is known as the $P_{k/h}$ function. A second metric used to quantify a component's vulnerability is A_v, the vulnerable area of the component. Component vulnerable area is defined as the presented area of the component that, if hit, would cause a kill of the component and is equal to the product of the component's presented area A_P in the threat approach direction and its $P_{k/h}$, i.e., $A_v = A_P \cdot P_{k/h}$.

The metrics used to quantify the vulnerability of the aircraft to a single random hit by a penetrator or contact-fuzed warhead include $P_{K/H}$, the probability the aircraft is killed given a random hit on the aircraft and A_V, the aircraft's single hit vulnerable area.[12] The metric used to quantify the vulnerability of an aircraft to the proximity- and time-fuzed HE warheads on AAA projectiles and guided missiles is $P_{K/D}$, the probability the aircraft is killed given an external detonation by a high-explosive warhead. The $P_{K/D}$ is a function of the location of the detonation point with respect to the aircraft.

WHAT ARE THE TWO METHODOLOGIES USED TO ASSESS VULNERABILITY?

In general, there are two methodologies used to assess aircraft vulnerability. One method is the a priori prediction of aircraft vulnerability by using analyses or modeling. This method is nearly always supported by prior live fire test data on component $P_{k/h}$ values for the various kill modes. However, the data have often been obtained on older equipment. The other method is the a posteriori observation and

---

[11] Other metrics sometimes used for component vulnerability are $P_{k/h}$, the probability a component is damaged given a hit, area removal, energy density, and blast.

[12] Lowercase subscripts refer to a component and uppercase subscripts refer to the aircraft. Thus, $P_{k/h}$ is the probability a component is killed given a random hit on the component, $P_{k/H}$ is the probability a component is killed given a random hit on the aircraft, and $P_{K/H}$ is the probability the aircraft is killed given a random hit on the aircraft.

possible measurement of aircraft vulnerability by using empirical data obtained from either actual combat, aircraft accidents, or controlled live fire testing.[13] This method is nearly always supported by a priori predictions of vulnerability prior to testing to define the test conditions and by a posteriori analyses or evaluation of the data. A brief review of the state-of-the-art of vulnerability analysis/modeling and vulnerability testing is given below.

Analysis/Modeling.

The prediction of an aircraft's vulnerability to the ballistic projectiles and guided missiles likely to be encountered in combat can be accomplished by using standardized computer programs.[14] One set of programs is applicable to a single hit by impacting penetrator or fragment. Computation of Vulnerable Area and Repair Time (COVART) is the Joint Technical Coordinating Group on Aircraft Survivability (JTCG/AS) standard program for computing the critical component vulnerable areas Av and the aircraft's vulnerable area AV for a single random hit by a penetrator or fragment (JTCG/ME, 1984). Another set of programs computes aircraft vulnerability to contact-fuzed HE warheads that detonate on the surface or within the aircraft. High Explosive Vulnerable Area and Repair Time (HEVART) (BRL, 1978 and HEI Vulnerability Assessment Model (HEIVAM) (Datatec Inc., 1979) are examples of this type of program. A third set, known as endgame programs, computes the probability an aircraft is killed due to an external burst of an HE warhead. SCAN (Dayton University Ohio Research Institute, 1976) is the current JTCG/AS endgame model for computing an aircraft's PK/D. Modular Endgame Computer Assessment (MECA), Joint Services Endgame Model (JSEM), SESTEM II (ASD/WPAFB, 1981), and SHAZAM (Air Force Armament Lab./Eglin AFB, 1983) are four other widely used endgame programs.

All of these vulnerability assessment programs require as input a three-dimensional data base that defines the geometric model of the aircraft. The geometric model may be contained within the vulnerability assessment program, as in SCAN, or it may be developed in a separate program, such as MAGIC, Ballistic Research Laboratory Computer-Aided Design (BRL-CAD) package, or FASTGEN III, which are used as preprocessors for COVART. This model should contain all of the aircraft's components, equipment, and supplies, including such items as fuel, hydraulic fluid, and ordnance. However, because of the limitations on program size, available time, and manpower, many small noncritical components that are not expected to influence the results are often omitted.[15] Another subsystem that has often been omitted in vulnerability assessments is the on-board ordnance in the form of bombs, missile warheads and propellants, and ammunition drums. On most aircraft, bombs and missiles are carried externally. In this position, they may shield other components from projectiles and fragments, or they may react violently to a ballistic impact (e.g., detonate) and destroy the aircraft. The new stealth aircraft carry ordnance internally in order to reduce signatures. Adverse reactions of any internally carried ordnance, such as a deflagration or a detonation, have an even greater probability of destroying the aircraft. The omission of on-board ordnance from the assessment is discussed in more detail in Chapters 2 and 4.

Another input requirement for the assessment is the kill tree (or logical kill expression) for the selected kill category (and level if appropriate). This tree defines the redundant and nonredundant components that if killed individually (the single engine on a single-engined aircraft) or in combination (both engines on a two-engined aircraft) will cause an aircraft kill. Associated with each critical component on the tree is a data base that contains the Pk/h or Av value for the component that is based upon the selected threat weapon or damage mechanism and the possible range of impact velocities on the installed component, for the kill modes considered in the critical component analysis.

Vulnerability to a Single Hit by a Penetrator or Fragment.

All of the vulnerability assessment programs contain an assumption as to how the damage mechanisms associated with the weapon proceed through the aircraft. The COVART methodology assumes that the penetrator or fragment from any selected direction[16] is equally likely to impact the aircraft at any location and that it propagates along a straight line, known as a shotline, through the aircraft, slowing down and possibly breaking up as it penetrates the various components. The amount of fragment or penetrator slowdown is determined by the penetration equations that are a part of the built-in data base. Ricochet of the fragment or penetrator is not considered. An additional assumption often made is that only the components that are intersected by one shotline can be killed by the hit along that shotline. This assumption rules out the possibility of cascading damage away from the shotline.[17] In the analysis, the presented area of the aircraft

---

[13] Combat and accident data are extremely valuable as adjuncts to the other methodologies, but they are limited in scope, limited in the information on the nature of the event, and not always available for direct application.

[14] The Joint Technical Coordinating Group on Aircraft Survivability has established a library of computer programs for assessing the susceptibility, vulnerability, and survivability of aircraft. The library is maintained and operated by the Survivability/Vulnerability Information and Analysis Center (SURVIAC) at the Wright Aeronautical Laboratories.

[15] The COVART model for the F-22 contains 2,213 components, of which nearly half are critical.

[16] The directions usually selected include the six cardinal views of front, back, top, bottom, left side, and right side, and may include the twenty 45-degree angles between these six views.

[17] It is possible to modify the intersected component's $P_{k/h}$ to account for kills of adjacent components.

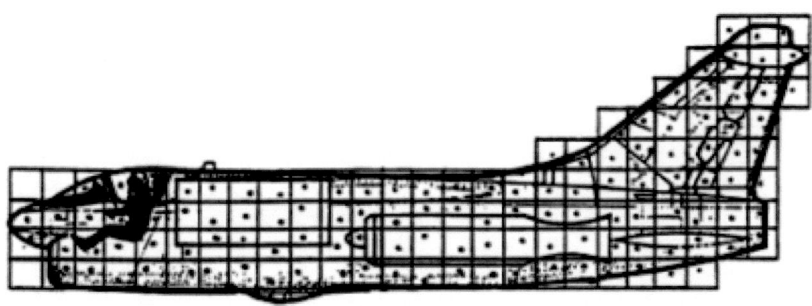

Figure 1-2 Example of a grid and random shotlines from FASTGEN for COVART (Ball, 1985). Copyright © AIAA 1985—Used with permission.

from the selected direction is covered by a uniform grid, and one shotline is randomly located within each cell. An example of the random shotlines within the cells for a particular aircraft is shown in Figure 1-2.

The user has the option of selecting the uniform cell size. Typical cell sizes range from 12 inches to 1 inch on a side, with 2 inches being typical. A preprocessor program, known as a shotline generator program, such as MAGIC, BRL-CAD, or FASTGEN III, identifies all of the critical components intersected by each shotline. This information is input data for COVART. COVART computes the vulnerable area of each critical component and the aircraft's single hit vulnerable area, as well as the probability the aircraft is killed by a random hit. For component vulnerable areas, each grid cell containing a shotline that intersects a component has a vulnerable area equal to the product of the presented area of the cell and the $P_{k/h}$ for the shotline through the component. The total vulnerable area of the component is the sum of the vulnerable areas of those cells with shotlines that intersect the component. For the aircraft vulnerable area A_V, each grid cell shown in Figure 1-2 contributes a vulnerable area equal to the product of the presented area of the cell and the probability the aircraft is killed by a hit along the shotline in that cell.[18] The total aircraft vulnerable area is equal to the sum of the vulnerable areas of each of the cells. Consequently, redundant components, if separated, that both are not intersected by one shotline, do not contribute to the aircraft's single hit vulnerable area for that shotline.[19] The $P_{K/H}$ for the aircraft is equal to the A_V of the aircraft divided by A_P, the aircraft's presented area from the selected direction.

Vulnerability to a Contact-Fuzed High-Explosive Warhead.

Essentially the same analytical procedure is followed for contact-fuzed high-explosive warheads. A geometric model of the aircraft, the kill tree, and the critical component $P_{k/h}$ or Av data are required. A grid is superimposed on the aircraft and a shotline is randomly located within each cell. The difference between this analysis for the contact-fuzed HE warhead and the analysis for the single penetrator or fragment is the fact that components in the vicinity of the shotline can be killed by the blast and fragments from the detonation of the HE warhead. Thus, redundant critical components that are relatively close together can be killed by a single hit, causing a kill of the aircraft. Figure 1-3 shows the grid cell and randomly located shotlines for this type of analysis. Note that in this figure the HE warhead detonation can cause a kill of both the fuel tank and the engine even though neither component was hit directly by the weapon.

Vulnerability to an Externally Detonating High-Explosive Warhead.

The analysis for the externally detonating HE warhead, shown in Figure 1-4, follows the same procedure used for the single penetrator or fragment, except that the fragment shotlines emanating from the external detonation are radial rather than parallel, and the aircraft can suffer multiple fragment impacts over its surface rather than a single hit. In addition, the blast from the detonation can kill the aircraft. The assessment of the kill of the aircraft by external blast is usually made independently from the fragment assessment. Three-dimensional blast contours around the aircraft are determined as a function of HE weight. Within a particular blast kill contour for particular explosive charge weight, a detonation of a warhead with that charge weight or larger will kill the aircraft.

Results from the Analyses.

The results or information obtained from an analytical assessment of aircraft vulnerability

---

[18] When more than one nonredundant critical component is intersected by a shotline, the probability the aircraft is killed is equal to the union of the component probabilities of kill.

[19] This is the result of the assumption that only those components intersected by the shotline can be killed. A modification of the $P_{k/h}$ value for a component can be made to allow a hit on one component to cause a kill of another component due to cascading damage.

APPENDIX D

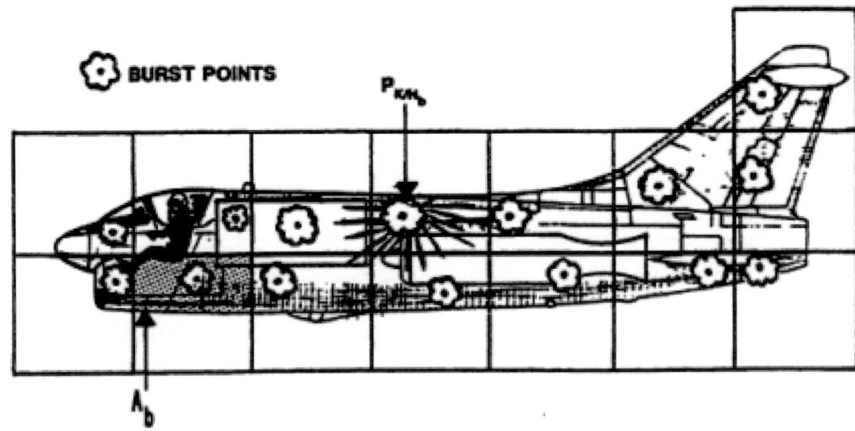

Figure 1-3 Grid cells and shotlines for the contact-fuzed high explosive weapon (Ball, 1985). Copyright © AIAA 1985—Used with permission.

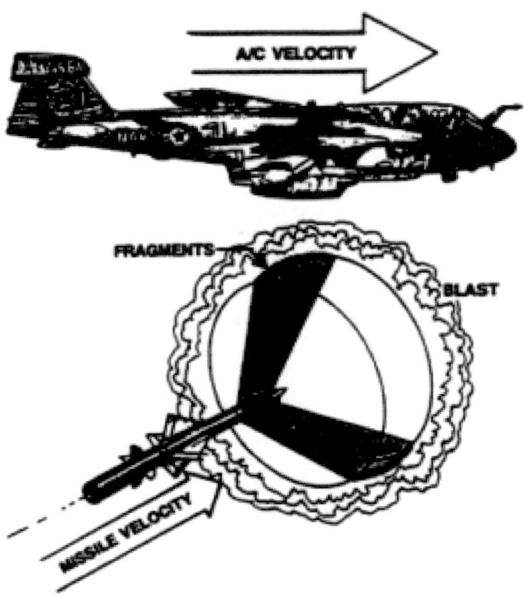

Figure 1-4 Aircraft vulnerability to the externally detonating HE warhead (Ball, 1985). Copyright © AIAA 1985—Used with permission.

for the single hit by a penetrator or fragment typically consists of predictions of the values of vulnerable area A_v for all of the critical components, the aircraft vulnerable area A_V, the probability the aircraft is killed given a hit within each grid cell, and the probability the aircraft is killed given a random hit $P_{K/H}$. The assessment results for the single hit by the contact-fuzed high-explosive warhead consist of the aircraft vulnerable area A_V and the probability of kill given a random hit on the aircraft $P_{K/H}$. The results of an assessment for the externally detonating warhead consist of the probability of kill of the critical components intersected by the fragment shotlines from the warhead detonation, the probability of aircraft kill due to blast, and the probability of aircraft kill given a detonation $P_{K/D}$.

REFERENCES

- Aeronautical Systems Division (ASD), 1981. Impacts of Engine Vulnerability Uncertainties on Aircraft Survivabilities, Wright-Patterson Air Force Base, Ohio, AD Number:C037839.
- Air Force Armament Laboratory, 1983. User Manual for the Air-to-Air Missile Program SHAZAM, Eglin Air Force Base, Fla., AD Number:B104959.
- Ball, R.E., 1985. The Fundamentals of Aircraft Survivability Analysis and Design, American Institute of Aeronautics and Astronautics, Inc., New York.
- Ballistic Research Laboratory (BRL), 1978. "HEVART-An Interim Simulation Program for the Computation of HEI Vulnerable Areas and Repair Times, Aberdeen Proving Ground, Md., AD Number:C030817L.
- Datatec Inc., 1979. High-Explosive Incendiary Vulnerability Model (HEIVAM), Volume 1, User Manual, Fort Walton Beach, Fla., AD Number:B107811L.
- Dayton University Ohio Research Institute, 1976. SCAN-A Computer Program for Survivability Analysis, Volume 1, User Manual, AD Number:B068149L.
- Deitz, P.H., et al., 1990. Current Simulation Methods in Military Systems Vulnerability Assessment, Ballistic Research Laboratory, Aberdeen Proving Ground, Md., BRL-MR-3880.
- Joint Technical Coordinating Group for Munitions Effectiveness (JTCG/ME), COVART II -A Simulation Program for Computation of Vulnerable Areas and Repair Times -Users Manual, 1984. Government Report Number:61 JTCG/ME 84-3.
- National Research Council (NRC), 1989. Armored Combat Vehicle Vulnerability to Anti-armor Weapons, A Review of the Army's Assessment Methodology, Committee on a Review of Army Vulnerability Assessment Methods, Board on Army Science and Technology, Commission on Engineering and Technical Systems, Washington, D.C.: National Academy Press.
- O'Bryon, James F., 1991. Presentation made to the Committee on Weapons Effects on Airborne Systems, July 24.
- U.S. Congress, 1986-1989. Survivability and Lethality Testing of Major Systems, DoD Authorization Acts, FY86—Sec. 123, FY87—Sec. 910—Sec. 910, FY88-89—Sec. 802.
- U.S. Congress, 1988. FY88-89 DoD Authorization Act Conference Report, Live-Fire Testing (Sec. 802).
- U. S. General Accounting Office (GAO), 1987. Live Fire Testing, Evaluating DOD's Programs, GAO/PEMD-87-17, Washington, D.C.: U.S. Government Printing Office.